Stuck In the Middle

Wars, Weapons, and the Forces That Will Shape the Next Thirty Years

Richard Lowe

The Writing King

Enemies of You Series

https://enemiesofyou.com

Stuck in the Middle

Table of Contents

See books by Richard Lowe at

https://masterofworlds.com

5

Get free publishing insights and industry updates at

https://thewritingking.substack.com

For ghostwriting and book coaching services see

https://thewritingking.com

Enemies of You Series

<https://masterofworlds.com/enemies-of-you>

The Death of Thinking
The Enslavement of Humanity

The Birth of the Augmented Human
The Freeing of Humanity

Turn Off The TV, Get Off Your Ass, and Do Something

Stuck in the Middle
Wars, Weapons, and the Forces That Will Shape the Next Thirty Years

The Enshittification of America
How Private Equity Destroyed the Things We Love

The Emasculation of America
How Russia's Long War Against the American Male Is Destroying the Nation From Within

The Villainization of America

———

See the full series description at the back of this book.

All books available at masterofworlds.com

Preface

Four books ago, I sat down to write about why Americans had stopped thinking. That turned into a book about why they'd stopped participating. That turned into a book about the technologies changing what humans can do. That turned into an investigation of how private equity gutted twelve industries people depend on to get through the day. And now here we are.

Each time, the research dragged me somewhere I didn't expect. The cognitive decline wasn't a cultural problem. It was producing citizens who couldn't govern themselves. The civic collapse wasn't laziness. It was a sane response to captured institutions. The technology story was really a geopolitical story about who builds the tools and who doesn't. And the private equity damage wasn't contained to hospitals and newspapers. It was hollowing out the country's ability to survive what's coming.

This book is what's coming.

Stuck in the Middle connects all four books into one argument. Cognitive decline feeds political dysfunction. Political dysfunction feeds military overreach. Military overreach feeds the great power competition. The competition collides with the structural forces reshaping the planet. And all of it points back to the technologies that could save us or destroy us, depending on who builds them and how they get deployed.

I'm not a foreign policy expert. I'm not a military analyst. I'm not an economist or a climate scientist or a demographer. I'm a writer who reads too much, asks questions that make people uncomfortable, and refuses to believe that complicated things can't be explained to normal people. Every expert I've met can explain their field to another expert. Almost none of them can explain it to a plumber, a teacher, or a retired electrician. That's my audience. People who are smart enough to understand the world but haven't been given the information in a form that respects their intelligence without requiring a PhD to decode it.

This book is written at a ninth-grade reading level. On purpose. Not because the ideas are simple. Because the audience deserves clarity. If you can't explain something clearly, you probably don't understand it yourself. I understand this material, and I've done my best to make sure you will too by the time you finish reading.

The book makes predictions. Some of them will be wrong. The Iran war broke out while I was writing this and reached a ceasefire before I finished, and the facts on the ground shifted daily the whole time. Casualty figures, oil prices, political developments will all be outdated by the time this reaches print. The structural analysis won't be. The missile math doesn't change because a ceasefire happens. The demographic cliff doesn't flatten because an election puts a new face on a screen. The Belt and Road doesn't stop because a summit produces a joint statement.

The forces are bigger than the events. This book is about the forces. Whatever shape the Iran war has taken by the time it reaches you, frozen or reignited or already a footnote, the structural arguments survive intact. The missile math is real whether the war is with Iran or a future adversary nobody has named yet. The industrial base hollowing is real whether the stress test is this war or the next one.

The scenario is a lens. The forces it reveals exist independent of the scenario. A reader in 2030 should read Parts One and Two as a stress test applied to the American military establishment, one that produces the same conclusions regardless of which specific conflict triggers it.

One more thing. This book is optimistic. That will surprise you after several hundred pages of disasters, failures, and structural threats. But the optimism is earned. Not the soft kind that says everything will work out. The hard kind that says the tools exist, the problems are solvable, and the only thing standing between the current mess and a dramatically better future is the decision to act. That decision is yours. Not a politician's. Not an expert's. Yours.

I'll see you on the other side.

Introduction: The Shape of the Trap

On February 28, 2026, the United States and Israel launched coordinated strikes against Iran. Within a week, the Strait of Hormuz was effectively closed, oil prices had spiked past $120 a barrel, American warships were burning through missile stockpiles faster than the defense industry could replace them, and a war that the majority of Americans opposed was consuming the military assets the country needs for a threat everyone agrees matters.

By the time the shooting stopped in April, the Supreme Leader was dead, his son had been installed by the Revolutionary Guard, and the Gulf states were calling Beijing instead of Washington. That situation, by itself, would be enough for a book. But the Iran war isn't the subject of this book. It's the entry point. The symptom that reveals the disease.

The disease is structural. The United States built the most powerful military in human history and then hollowed out the industrial base that sustains it. It built the world's greatest innovation engine and then defunded the research that feeds it. It built a global alliance system and then withdrew from it so gradually that nobody noticed until the system started breaking. It invited the world's best and brightest to come build their futures here and then spent two decades trying to slam the door shut. It produced the technologies that could solve every major crisis on the horizon and then refused to deploy them because the politics were too hard.

Meanwhile, on the other side of the world, a country with an older civilization and a longer time horizon has been watching, taking notes, and building. Not a military to match America's. An empire. Railroads instead of aircraft carriers.

Loans instead of missiles. Infrastructure instead of ideology. The largest construction project in human history, spanning three continents, creating dependencies that convert economic relationships into political control without firing a single shot.

This book is about the space between those two stories. The space where most of the world's people live. The space where the forces that will shape the next fifty years, demographic collapse, resource depletion, climate change, pandemic risk, technological revolution, and the great power competition, are all converging at once.

You are in that space. You are stuck in the middle.

. . .

The book is organized in seven parts, and each one builds on the ones before it.

Part One, *The Guns Don't Work*, is about the American military's stress test. The Iran war as a live demonstration of what happens when a force designed for short, cheap, technology-driven wars encounters an adversary with cheap mass-produced weapons and the production math to sustain a long fight. Missiles, ships, carriers, and the industrial base that is failing to support them.

Part Two, *Wars Without Reasons*, asks why the US fought a war that most Americans opposed, couldn't explain, and believed was making the country less safe. The shifting justifications, the ally whose objectives diverged from ours, the damaged nonproliferation regime, and the outcome no one in Washington planned for: a more hardline Iran controlled by the Revolutionary Guard, and a Middle East where China is the power the Gulf states now call.

Part Three, *The Long Retreat*, zooms out to the thirty-five-year withdrawal from global leadership that started before

11

most Americans were born, the economic forces driving it, and the world war that is already underway in a shape most people don't recognize.

Part Four, *The Four Horsemen*, describes the structural forces that are bigger than any war or any president: demographic collapse, the oil-fertilizer-food chain, climate change, and the next pandemic. Each interacts with the others to create compound effects worse than the sum of their parts.

Part Five, *The Quiet Empire*, maps what China is doing while the US stares at Taiwan. The Belt and Road as strategic colonization. The grand bargain for the Russian Far East. The construction of an alternative world order that doesn't depend on American permission or American institutions.

Part Six, *The Hollowed-Out Country and the Tools to Rebuild It*, starts with the damage, private equity extraction and the education crisis, and then lays out the technologies that could fix everything: nuclear energy, artificial intelligence, longevity science, genetic engineering, human enhancement, the innovation thesis, and space. Every one of these exists. None are deployed at scale.

Part Seven, *Stuck in the Middle*, assembles the model. Two hemispheres stress-tested against the forces. The India question. The better overlord. The migration century. And the question the book has been building toward: what are you going to do about it?

• • •

A note on sources and methodology. This book draws on publicly available information: government reports, academic research, news coverage, polling data, defense analysis, and the work of geopolitical thinkers, George Friedman chief among them, who have shaped my understanding of how structural forces interact with political decisions. Where I rely heavily on a specific analyst's framework, I credit them. Where I've synthesized multiple sources into an argument that is my own, I take responsibility for it. The conclusions are mine. The errors, if any, are mine. The opinions are definitely mine.

I wrote this book because I'm angry. Angry that the information in these pages is available to anyone who looks for it and yet almost nobody in a position of authority acts on it. Angry that the tools to solve these problems exist and gather dust while the political system fights about things that don't matter. Angry that my country, which has every advantage a civilization could ask for, is squandering those advantages through a combination of greed, ignorance, cowardice, and inertia.

But anger without direction is just noise. This book is the direction. Read it. Understand it. And then do something about it.

The trap has a shape. Understanding the shape is the first step to getting out.

Part One: The Guns Don't Work

Chapter 1: Where Are All the Guns?

When the Ukraine war broke out I kept waiting for someone in Washington to treat it as the canary in the coal mine. Here was a conventional war burning through ammunition at a rate nobody had planned for. Here was proof, in real time, that stockpile assumptions were wrong. I thought: someone will notice this and fix it. That did not happen. So now we are running low, and the implications of that are serious in ways that will matter when they matter most, which is during a crisis. I hope it gets resolved before then. I am not confident it will.

Pull up a photograph of the USS Missouri. The battleship that hosted the Japanese surrender in Tokyo Bay, September 1945. Look at her.

Nine sixteen-inch guns in three triple turrets, each barrel sixty-eight feet long and capable of throwing a shell the weight of a Volkswagen Beetle twenty miles. Twenty five-inch guns for medium range. Dozens of forty-millimeter anti-aircraft guns bristling from every surface like porcupine quills. Radar arrays, fire control directors, spotting towers. The ship looked like what it was: a floating fortress built to kill anything within line of sight and survive anything thrown back at it.

That was 1945. Eighty years of technological progress later, pull up a photograph of the USS Zumwalt. The most advanced surface warship the United States Navy has ever built. Four and a half billion dollars per copy.

She looks like a floating iPhone. Smooth gray surfaces angled to deflect radar. Two gun turrets forward, and neither one has working ammunition because the rounds cost $800,000 each and the Navy cancelled the program. The ship is so automated it carries a crew of 147, compared to Missouri's 2,700. From a distance she looks unarmed. From up close she still looks unarmed.

The first time most people see a modern warship, they ask the same question. Where are all the guns?

It's a reasonable question. And the answer tells you almost everything you need to know about how the American military got itself into the mess it's in right now.

. . .

The guns went inside. That's the short version.

Starting in the 1960s and accelerating through the 1980s, the Navy replaced visible gun turrets with something called the Vertical Launch System. VLS for short. Picture a grid of steel tubes sunk into the deck of a ship, each one holding a missile standing on its tail like a bullet in a revolver. The tubes are covered with flat hatches that sit flush with the deck. From the outside, you see nothing. No barrels, no turrets, no obvious weapons at all. Just smooth metal and a few antennas.

An Arleigh Burke-class destroyer, the workhorse of the current fleet, carries 90 to 96 of these cells. Each one can hold a Tomahawk cruise missile for hitting land targets at a thousand miles. Or an SM-2 for shooting down aircraft. Or an SM-6 for longer-range air defense. Or an SM-3 for killing ballistic missiles in space. Or an Evolved Sea Sparrow packed four to a cell for close-in defense. The ship's single visible weapon is a five-inch gun on the bow that handles the odd jobs.

On paper, this is a massive upgrade from the Missouri's gun-based arsenal. The Missouri could hit a target twenty miles away if the weather was good and the fire control officers were having a good day. A Burke can put a Tomahawk through a window at a thousand miles. The lethality per ton isn't in the same universe.

And there's a strategic logic behind the smooth surfaces. Every protruding turret, antenna, and gun barrel reflects radar energy. In the age of anti-ship missiles, your radar cross section is your life expectancy. A ship that's hard to find on radar is hard to target. A ship with a forest of gun barrels sticking up is a ship asking to get hit. The clean lines of a modern warship aren't aesthetic. They're survival.

So far, so smart. Fewer visible weapons, more actual lethality, better survivability. The logic made sense in every briefing room and every budget hearing for fifty years.

Then the Iran war started. And the logic broke.

. . .

On February 28, 2026, the United States and Israel launched coordinated strikes against Iran. What was supposed to be a short, sharp demonstration of overwhelming force turned into the first sustained naval conflict the US had fought since World War II. Iran fired back. Not with the kind of pinprick attacks the Pentagon had war-gamed against, but with everything it had.

In the first five days, Iran launched over 500 ballistic and naval missiles and nearly 2,000 drones. Not at American ships directly, because Iran couldn't reliably find them. At American bases across the Middle East, at Gulf state allies, at Israel, at anything associated with the coalition. The sky over the Persian Gulf, the Arabian Sea, and the eastern Mediterranean filled with incoming threats around the clock.

The ships did their job. Aegis cruisers and destroyers tracked incoming missiles and knocked them down. THAAD batteries ashore swatted ballistic missiles at high altitude. Patriot batteries caught what leaked through lower down. The layered defense system worked exactly as designed.

For about a week.

Then the numbers started telling a different story. Each THAAD interceptor costs roughly twelve million dollars. Each SM-6 runs four to five million. Iran was building the ballistic missiles those interceptors were designed to kill for a few hundred thousand dollars apiece. And Iran could build about ninety missiles a month. The US could manufacture six or seven interceptors in the same period.

Read those numbers again. Ninety versus seven. That's not a production gap. That's a production chasm.

Four days into the war, at least one Gulf ally was already running low on interceptors. Four days. Not four months. The United States burned through approximately a quarter of its entire stockpile of THAAD interceptors defending Israel during the opening phase. A system that took years to manufacture was being consumed in days.

The Houthis had been sending this same message for two years and nobody listened. American destroyers in the Red Sea had been firing million-dollar SM-2 missiles to shoot down drones that cost twenty thousand dollars. Do that math enough times and you go bankrupt. The Iran war was the Houthi problem at industrial scale, and nobody had done anything about it in the two years between the warning and the test.

• • •

Here's what the missile math reveals about the deeper problem. The US military doesn't have a wartime industrial base. It has a peacetime industrial base that panics during wars.

The factories that make interceptor missiles weren't built for surge production. They were built to fill peacetime orders at peacetime rates, generating peacetime profits for defense contractors whose stock prices depend on steady, predictable revenue, not on the ability to triple output in an emergency. The skilled workers who assemble rocket motors and precision seekers aren't sitting in a reserve pool waiting to be called up. They're employed at capacity. The sub-tier suppliers who make the specialized components have been through rounds of consolidation and private equity ownership that stripped out redundancy and excess capacity because redundancy doesn't look good on a quarterly earnings report.

When the shooting started and the Pentagon scrambled to ramp up production, the money hit organizations that couldn't absorb it quickly. Lockheed signed a deal to quadruple THAAD interceptor output from 96 to 400 per year. Raytheon committed to ramping up Tomahawks, SM-6s, and other systems. Those are impressive numbers on a press release. They're meaningless in a war that's consuming the stockpile faster than the ramp-up can replace it.

And here's the part that should keep you up at night: when this war ends, the pattern will repeat. The panic spending will slow. Budget hawks will argue that the war is over and we don't need to maintain wartime production rates. The lines will cool. The specialized workers will drift to other industries. The sub-tier suppliers will get bought by private equity firms looking for the next extraction opportunity. And fifteen years from now, when the next crisis hits, someone in the Pentagon will discover that the cupboard is bare. Again.

This has happened after every American war since Korea. The cycle is so predictable it should have its own name. Build-up, burn-through, draw-down, atrophy, panic. Repeat.

• • •

But the missile math is only half the story. The other half is what those missiles are protecting.

The US Navy has 293 battle force ships. That's the official count as of late 2025. In the Reagan era, the fleet numbered 594. During World War II it peaked at 6,768. Today's Navy is the smallest it's been in over a century, and it's expected to get even smaller. The Congressional Budget Office projected the fleet would hit a low of 283 ships in 2027 because the Navy is retiring ships faster than it's commissioning new ones.

Let that sink in. The fleet is shrinking during an active shooting war.

Congress mandated a 355-ship Navy back in 2018. The Navy later raised its own target to 381. The current administration has something called the "Golden Fleet" goal that it hasn't even published the details of. None of these numbers have been met. They are PowerPoint aspirations backed by shipyard capacity that doesn't exist.

America has four major shipyards that build combatant vessels. All of them are behind schedule on current projects. The Columbia-class submarine program, which is building the replacement for America's nuclear missile boats, is consuming a huge share of available capacity. You can't just stand up a new shipyard the way you can add a missile production line. We're talking a decade to build one, train the workforce, and get it certified.

Workforce. That's the word that keeps coming up when you dig into why things aren't getting built. Not funding. Not technology. Not design. Workforce. There aren't enough welders, pipefitters, electricians, and nuclear-qualified technicians to build the ships the Navy says it needs. The vocational training pipeline that once produced these workers was dismantled over the past forty years under the assumption that everyone should go to college instead. Germany kept its apprenticeship system and can build things. The United States sent its kids to get communications degrees and now can't find people who know how to weld.

At Newport News Shipbuilding, the nation's sole builder of nuclear aircraft carriers, a pipe welder named Wobser earned a reputation for having one of the lowest X-ray reject rates in the yard. Every nuclear weld gets X-rayed to check for flaws. An imperfect weld on a submarine reactor system isn't a quality issue. It's a catastrophe waiting to happen. Wobser, self-taught before the yard's welding school refined his technique, became one of the mentors who trained the next generation of nuclear welders. Those mentors are the ones who carry the institutional knowledge that keeps the production line running, the ones who can feel when a weld isn't right before the X-ray confirms it.

When COVID hit in 2020, a wave of workers like Wobser retired. Xavier Beale, the yard's vice president of human resources, told Defense News that replacing master tradespeople with recent high school graduates had measurably affected productivity. The yard needs to hire 19,000 skilled tradespeople over the next decade. It's busing workers in from North Carolina because it can't find enough in the Hampton Roads region.

There are roughly 230 miles of pipe on a nuclear carrier. Every inch of it has to be fabricated, fitted, welded, inspected, and tested by people who know what they're doing. You can't automate that. You can't outsource it. And you can't train a replacement in a weekend.

So the ships can't be built fast enough. And the ships that do exist are being run into the ground.

• • •

During the Iran war, forty percent of available US aircraft carriers were deployed to support one conflict. That's not a typo. Four out of ten carriers in one theater, in a Navy that's supposed to maintain global presence across seven oceans.

The USS Gerald R. Ford, the newest and most expensive carrier in the fleet at $13.3 billion, had been deployed for over 300 days when a fire broke out in its laundry area on March 12. Not a combat hit. A laundry fire. The fire burned for over 30 hours before it was fully contained. More than 600 sailors lost their berthing. The Navy had to strip 1,000 mattresses off the future USS John F. Kennedy, still under construction at Newport News, and fly them to the Ford. Two thousand sweatsuits were collected and shipped because the crew couldn't wash their clothes. Senator Tim Kaine of Virginia wrote to the Secretary of the Navy warning that the ship's crew was being "pushed to a breaking point." The ship, which had already been dealing with a broken sewage system and other mechanical problems, withdrew from operations and limped to Crete for repairs.

The most advanced warship ever built, knocked out of a war by its own laundry room.

Meanwhile, the USS John C. Stennis was fourteen months behind schedule and half a billion dollars over budget on its mid-life overhaul. The USS Harry S. Truman was entering the shipyard for its own five-year overhaul. The USS Nimitz, which was supposed to be decommissioned, had its service life extended because there simply weren't enough carriers to go around.

Every carrier in the Middle East is a carrier that's not in the Pacific. Every destroyer burning through its VLS magazine defending Gulf state airspace is a destroyer that's not watching the Taiwan Strait. The Navy designed to deter China is being consumed by a war against Iran, a country with a GDP smaller than Italy's.

* * *

This is where the gun question circles back to something bigger than naval architecture.

The American military was designed around a set of assumptions that all sounded reasonable in the briefing room. Wars will be short. Precision weapons will keep costs manageable. Technology will substitute for mass. Quality will beat quantity. We'll fight one major war at a time and maintain deterrence elsewhere with forward presence. The industrial base will surge when needed.

Every one of those assumptions is being tested right now, and most of them are failing.

Wars are not short when your adversary has enough missiles and drones to keep shooting for months. Precision weapons are not cost-effective when the other side's weapons cost a hundredth of your interceptors. Technology does not substitute for mass when you run out of missiles and can't make more fast enough. Quality does not beat quantity when quantity means you physically can't defend everywhere you need to defend. And the industrial base does not surge because the industrial base was hollowed out by decades of financial optimization that treated surge capacity as waste.

The warships look naked because the military believed it could replace visible, brute-force firepower with invisible, precise, technology-driven lethality. In a short war against a weak enemy, that trade works. In a long war against a determined enemy with cheap mass-produced weapons, it doesn't. You run out of expensive things before they run out of cheap things. That's not a strategy. That's a math problem, and the math doesn't care about your technology.

The Navy isn't alone in this. The entire American military establishment, from the Army's precision munitions to the Air Force's stealth aircraft, was built on the same set of assumptions. Short wars. Precision. Technology. Quality over quantity. Small but lethal. The Iran war is stress-testing all of it.

And China is watching. Taking notes. Counting every missile expended, every carrier sidelined, every production bottleneck exposed. Not because China plans to fight the same kind of war Iran is fighting. Because China plans to fight a much bigger one, and it wants to know exactly how fast the American magazine runs dry.

• • •

None of this is unsolvable. Directed energy weapons, things like high-powered lasers and microwave systems, can kill drones and missiles at the cost of diesel fuel and electricity rather than million-dollar interceptors. Autonomous drone swarms can provide the cheap mass that expensive precision weapons can't. Electromagnetic rail guns could fire projectiles at a fraction of missile costs. The technology for all of this exists in various stages of development. The Navy has been talking about laser weapons for twenty years.

Talking about them. Not deploying them.

Bryan Clark spent twenty years in the Navy as a submariner, then served as special assistant to the Chief of Naval Operations, helping write Navy strategy. He's now at the Hudson Institute, and he's been saying the quiet part out loud for years. The Navy's surface fleet, Clark argues, was designed around the assumption that it would fight short engagements where magazine depth didn't matter. Ninety-six VLS cells per destroyer sounded like a lot when your war plan assumed a week of fighting followed by a port visit to reload. Against an adversary who can sustain missile and drone attacks for months, ninety-six cells is a countdown to empty.

Clark testified before Congress that the Navy would lose roughly 788 VLS cells by 2027 just from retiring its Ticonderoga cruisers, each of which carried 122 cells. That's the equivalent of eight destroyers worth of firepower walking out the door, with nothing coming in fast enough to replace it. His proposed fix is blunt: stop building fewer and fewer exquisite ships and start building more platforms that can carry missiles, including unmanned ones, so the fleet has enough magazine depth to sustain a real fight.

Then-Chief of Naval Operations Admiral Mike Gilday said the same thing in 2019: "We can't continue to wrap $2 billion ships around 96 missile tubes in the numbers we need to fight in a distributed way against a potential adversary that is producing capability and platforms at a very high rate of speed." The Navy's response has been to keep ordering the same ships it was already building. Meanwhile, by the end of 2024, China's fleet had surpassed 50 percent of the US Navy's total VLS capacity on surface ships, up from less than 1.5 percent in 2005.

The institutional inertia of a military built around expensive, exquisite, small-batch weapons systems is enormous. The contractors who build those systems have no incentive to develop cheap alternatives that would cannibalize their revenue. The admirals who rose through ranks defined by carrier battle groups have no incentive to champion a force architecture that doesn't center on carriers. The Congress members whose districts depend on current production lines have no incentive to redirect funding to new technologies that would be built somewhere else.

So the Navy that fought the Iran war in 2026 looks almost exactly like the Navy that was designed in the 1990s. Thirty years of technological change, and the force architecture barely moved. Not because better alternatives don't exist. Because the system that decides what gets built is optimized to build what it's already building.

Sound familiar? If you've read *The Enshittification of America*, you've seen this pattern before. An institution captured by financial incentives that optimize for the wrong things, producing outcomes that serve the people running the system while failing the people the system is supposed to protect. The defense industrial base isn't being eaten by private equity in exactly the same way as your local hospital or newspaper. But the underlying logic is identical: short-term extraction beats long-term investment, every time, until the system hits a wall.

The Iran war is the wall.

The warships look undergunned because the military traded mass for precision, volume for technology, and resilience for efficiency. Those trades made perfect sense in a world of short, cheap wars against overmatched opponents. We don't live in that world anymore. The question now is whether the United States can adapt before the next war, the bigger one, the one that everyone in the Pentagon knows is coming, arrives.

Because the next one won't be fought against a country with the GDP of Italy.

It'll be fought against the country that's been watching this whole time, counting our missiles, and building railroads.

Chapter 2: The Missile Math

There's a factory in Troy, Alabama, where Lockheed Martin builds Javelin anti-tank missiles. A few years ago, they could produce about 2,400 a year. They're trying to ramp that up to nearly 4,000. The Javelin is a relatively simple weapon by modern standards. It's a tube with a seeker head, a rocket motor, and an explosive charge. It costs about $175,000 per shot.

Now try to build something harder. A THAAD interceptor has to fly into the upper atmosphere, detect a ballistic missile warhead traveling at several times the speed of sound, calculate an intercept trajectory in real time, and hit it with the kinetic force of a head-on car crash at Mach 8. No explosive warhead. Just speed and precision. A bullet hitting a bullet, except both bullets are moving faster than any gun ever made.

That costs twelve million dollars. And the United States can produce about 96 of them a year.

Ninety-six. In a year. To defend the entire country and all of its allies worldwide.

Iran can build the ballistic missiles that THAAD is designed to kill at a rate of roughly ninety per month. Those missiles aren't as sophisticated. They don't need to be. They just need to fly in roughly the right direction, get past the defenses, and blow up near something valuable. Their guidance systems use Chinese BeiDou satellite navigation instead of GPS, so the US can't jam them easily. They cost maybe $200,000 to $300,000 each.

This is the arithmetic that the American defense establishment spent thirty years ignoring. Not because nobody did the math. Because the math was inconvenient.

The problem has a name in military circles: cost-exchange ratio. It means the relationship between what the attacker spends and what the defender spends to counter it. In a functional defense system, you want that ratio to favor the defender. You want your defenses to be cheaper than their weapons, so that a sustained campaign bleeds the attacker dry while you remain solvent.

The US missile defense architecture has the ratio backwards. Catastrophically backwards.

A THAAD interceptor at twelve million dollars to kill a ballistic missile that costs three hundred thousand dollars is a cost-exchange ratio of 40 to 1 in favor of the attacker. An SM-6 at five million dollars to kill a cruise missile that costs half a million is 10 to 1 in the attacker's favor. An SM-2 at two million dollars to shoot down a drone that costs twenty thousand is 100 to 1.

One hundred to one.

These aren't obscure numbers from a classified briefing. Analysts have been writing about cost-exchange ratios in open publications for a decade. The Houthi campaign in the Red Sea turned it into a live demonstration. American destroyers were firing their way through their VLS magazines shooting down drones that the Houthis were building in garages with Iranian parts. Each engagement was a financial victory for the guys in sandals and a step closer to empty tubes for the guys with the billion-dollar warships.

The Iran war took that dynamic and multiplied it by a factor of fifty. Not garage drones picking at a few ships. Hundreds of ballistic missiles, thousands of drones, aimed at targets across half a dozen countries at the same time. The defenders had to intercept nearly everything because the targets were cities, military bases, oil infrastructure, and allied nations who expected American protection. You can't choose to let a ballistic missile through because it's headed for Abu Dhabi instead of Tel Aviv. You shoot at all of them.

And each time you pull the trigger, you burn through a weapon you can't replace for months.

. . .

The production bottleneck isn't money. Congress can appropriate funds overnight. The bottleneck is physics, metallurgy, and people.

A modern interceptor missile is one of the most complex manufactured objects on the planet. The solid rocket motor requires propellant mixed to exact specifications and cured at precise temperatures. The seeker head uses infrared or radar components that require clean-room assembly. The guidance electronics use chips and processors that come from a supply chain stretching across multiple countries. The airframe has to withstand stresses that would tear apart an airliner. Every component has to work perfectly, the first time, every time, because there's no second chance when you're hitting a warhead at Mach 8.

The people who build these things are specialists. You don't train a rocket motor technician in a weekend seminar. The institutional knowledge that keeps a production line running, the tribal knowledge of how to handle materials that don't behave according to the textbook, the experienced hands that can feel when something isn't right, takes years to develop. When a production line shuts down and those workers disperse, that knowledge walks out the door with them. When the line restarts five or ten years later, you're not resuming production. You're rebuilding a capability from something close to scratch.

Northrop Grumman, which makes the solid rocket motors used in multiple missile systems, invested a billion dollars over six years to expand capacity. After all that spending, they produce five to six thousand motors a year for the GMLRS rocket system alone. That sounds like a big number until you realize that a single HIMARS launcher can burn through a pod of six rockets in a few minutes. The Ukraine war chewed through thousands. The Iran war is chewing through more. The motor factory is the valve that controls how fast the entire system can produce, and the valve doesn't open any wider just because someone writes a bigger check.

Now multiply that bottleneck across every component in every missile system. Seekers. Guidance boards. Fins. Wiring harnesses. Thermal batteries. Each one has its own supply chain, its own specialized manufacturers, its own lead times. Many of those manufacturers are small companies, sole-source providers, the kind of operation where one key engineer retiring can stall production for months. These aren't companies that show up on CNBC. They're machine shops in the Midwest and electronics fabricators in New England, and half of them have been through private equity ownership cycles that stripped out every dollar of excess capacity because excess capacity is inefficiency and inefficiency is the enemy of quarterly returns.

• • •

In January 2026, a month before the war started, Lockheed Martin signed a landmark deal to quadruple THAAD interceptor production from 96 to 400 per year. They broke ground on a new "Munitions Acceleration Center" in Camden, Arkansas. A separate agreement tripled PAC-3 interceptor capacity from roughly 600 to 2,000 annually over seven years.

Seven years.

Raytheon committed to ramping up to over 1,000 Tomahawk cruise missiles per year, around 1,900 AMRAAMs, and roughly 500 SM-6 interceptors. These are all multiples of previous production rates. On paper, they look like the kind of industrial mobilization that wins wars.

In practice, they're capacity targets for the end of a multi-year ramp. They don't help you in March 2026 when you're burning through your existing stockpile at a rate that will empty it before the new production comes online. It's like having your house on fire and signing a contract to install a sprinkler system next year.

President Trump posted that the US has a "virtually unlimited supply" of mid-grade munitions. He was talking about bombs and rockets, the bread-and-butter stuff that aircraft drop on buildings. He wasn't wrong about those. But when it came to the high-end interceptors, the ones that stop ballistic missiles and protect cities and bases, he admitted the stockpile was "not where we want to be."

Not where we want to be. During an active war. That's a notable admission from a president who doesn't admit weakness easily.

• • •

The historical pattern is what makes this maddening rather than merely alarming.

After World War II, the US demobilized at a speed that stunned even its own generals. Twelve million men and women in uniform dropped to 1.5 million within two years. Factories that had been producing tanks and aircraft converted back to cars and refrigerators. Then Korea hit in 1950 and the military scrambled to reconstitute a force it had just dismantled.

After Korea, a drawdown. After Vietnam, a bigger drawdown. The "hollow Army" of the late 1970s became a punchline about military readiness. Reagan rebuilt it, won the Cold War, and then the drawdown started again. Clinton's "peace dividend" cut the military by a third. The defense industrial base consolidated from dozens of major contractors to five or six. Production lines closed. Skilled workers retired or moved on.

Then 9/11 hit and the system scrambled again. Massive spending on precision munitions for Iraq and Afghanistan. Production lines reopened. Workers retrained. By the time those wars wound down, the pattern reasserted itself. Sequestration in 2013, triggered by the Budget Control Act's failure to produce a bipartisan deficit deal, slashed $46 billion from the defense budget in a single year, with $454 billion in total defense cuts mandated over the following decade. The cuts were across-the-board and indiscriminate, hitting readiness accounts hardest. Defense Secretary Leon Panetta called it "legislative madness." The Joint Chiefs of Staff signed a joint letter warning that readiness was "at a tipping point" and the military was "on the brink of creating a hollow force." Hundreds of thousands of civilian Defense Department employees had their work weeks cut to four days. Production rates dropped. The supply chain atrophied again.

The Ukraine war was supposed to be the wake-up call. Watching a European land war consume artillery shells at World War I rates while Western stockpiles drained within months should have triggered an industrial mobilization. It triggered press releases and contract announcements. The actual shells took years to arrive. One former undersecretary of defense admitted in 2023 that the US had "allowed production lines to go cold, watched as parts became obsolete and seen sub-tier suppliers consolidate or go out of business entirely."

That was the situation going *into* the Iran war. Not after decades of further neglect. That was the starting position, already degraded, already behind, already short.

And now the same officials who let the production lines go cold are appearing on cable news expressing concern about stockpile levels. The same contractors who optimized for steady peacetime profits are announcing heroic ramp-up plans that won't deliver at scale for years. The same Congress that funded sequestration is holding hearings about why we don't have enough missiles.

The cycle isn't a mystery. It's a choice. A choice made by defense contractors who prefer predictable revenue over surge capacity. A choice made by Congress members who prefer cutting military budgets during peacetime to get money for domestic priorities. A choice made by Pentagon leaders who prefer investing in the next generation of exquisite weapons over boring production infrastructure for current ones. A choice made by private equity firms who buy sub-tier defense suppliers, strip the excess capacity, extract the value, and leave the hollowed-out shell for the next war to discover.

Every link in the chain is acting rationally within its own incentive structure. And the collective result is a defense industrial base that cannot sustain the country in a real war.

● ● ●

There's a deeper problem buried in the missile math that goes beyond production rates and stockpile levels. It's about what kind of wars the United States is prepared to fight versus what kind of wars the United States is fighting.

The precision revolution that started in the 1991 Gulf War convinced an entire generation of military planners that technology had changed the nature of warfare. One smart bomb could do what used to require a hundred dumb ones.

One cruise missile could hit a specific building from a thousand miles away, a task that previously required a squadron of bombers and a prayer. The footage of precision strikes going through ventilation shafts became the defining image of American military power.

And it worked. Against Iraq. Against Serbia. Against Libya. Against the Taliban in the opening weeks of Afghanistan. Every opponent who stood up and tried to fight the United States in conventional battle got demolished by precision weapons delivered from standoff range. The lesson seemed clear: you don't need mass. You need accuracy.

But Iraq, Serbia, Libya, and the Taliban had something in common. They were all short wars against opponents who couldn't shoot back effectively. None of them could threaten American bases with ballistic missiles. None of them could saturate air defenses with hundreds of simultaneous threats. None of them had an industrial base capable of producing weapons faster than the US could intercept them.

Iran can. Not as well as China could, but well enough to expose the flaw in the entire theory.

The flaw is this: precision weapons are tools for short wars. In a short war, you have enough. You expend a fraction of your stockpile, achieve your objectives, and go home to rebuild. In a long war, you need volume. You need weapons cheap enough to use in quantity and a production base capable of replacing them as fast as you fire them. The US has optimized entirely for precision and completely neglected volume. It can kill anything it can see. It just can't keep doing it for more than a few weeks.

Iran figured this out. So did the Houthis. And China has been building its military doctrine around this insight for twenty years. Their approach is simple: force the US to defend against threats so numerous and so cheap that the defense costs more than the attack. Don't try to match American precision. Overwhelm American capacity.

It's not a new idea. It's the oldest idea in warfare. Mass beats quality when mass is cheap enough and quality is expensive enough. The Persians threw their army at the Spartans at Thermopylae on the same principle. The Soviets drowned the Wehrmacht in T-34s on the same principle. The only thing that's changed is the arithmetic. A hundred-to-one cost-exchange ratio means that a regional power with a fraction of America's GDP can bleed the American military dry in a sustained conflict.

That's what the missile math shows. Not that American weapons don't work. They work brilliantly. Every interceptor that hit an Iranian missile was a triumph of engineering. The problem is that the triumph costs fifty times what the threat costs, you can't make them fast enough to keep up, and the other side knows it.

• • •

So what do you do?

The answer is you change the math. You have to develop weapons that cost less to fire than the things they're shooting down, that can be produced in volume, and that don't depend on supply chains that take years to ramp up.

High-energy lasers can shoot down drones and rockets at the cost of diesel fuel for the generator. The Navy has tested them. The Army has prototypes. They work against the cheap threats that are draining the expensive interceptors. A laser doesn't run out of ammunition as long as the ship has power. It doesn't cost a million dollars per shot. And it doesn't depend on a rocket motor factory in Alabama that needs seven years to triple capacity.

Autonomous drone swarms can provide the mass that precision weapons can't. The Pentagon's Replicator program, launched in 2023 with $1 billion in funding, aimed to field thousands of cheap, expendable drones by 2025. It delivered hundreds. The concept is sound: small, simple, numerous, and lethal enough that any enemy has to deal with them. The cost-exchange ratio flips when you're the one sending cheap things and they're the ones spending expensive interceptors. But even a program designed to move fast got bogged down in the acquisition system it was supposed to bypass.

These solutions exist. Some of them are in testing. Some are in limited deployment. None of them are at the scale needed to change the equation in the current war, because the current war arrived before the solutions did.

That's the story of the missile math. The US built the world's most lethal military, weapon for weapon, system for system, dollar for dollar. And then it ran into an adversary who didn't need to match that lethality. Who just needed enough cheap weapons to make the math work in his favor.

The question for the next chapter isn't whether the US can build better weapons. It can. The question is whether it has enough ships to put them on. And the answer to that question is the one the Navy doesn't want to talk about.

Chapter 3: The Incredible Shrinking Navy

Disgust. That is the honest word. We had the largest, most capable navy in the world. We still have a great navy. But it cannot fight a long war and it does not have the ship mix it needs. The numbers tell the story: from nearly 600 ships down to around 290, with maintenance backlogs so severe that a significant portion of what we have cannot deploy. We inherited something extraordinary and let it erode through neglect, budget games, and the procurement dysfunction this chapter documents. That is not a strategic choice. It is a slow institutional failure that nobody wanted to own.

In 1987, the United States Navy had 594 ships. Ronald Reagan had spent most of a decade rebuilding the fleet after the post-Vietnam hollowing, and the result was a force that could credibly threaten to fight the Soviet Navy on every ocean at the same time. Fifteen carrier battle groups. Nearly a hundred attack submarines. Battleships pulled out of mothballs and given Tomahawk missiles. Cruisers, destroyers, frigates, and amphibious ships in numbers that let the Navy do what navies have always done: be everywhere at once, or at least make your enemies believe you might be.

Today the US Navy has 293 battle force ships. It hasn't been above 300 since 2003. The fleet that won the Cold War has been cut in half, and the half that's left is older, more overworked, and more thinly spread than at any point in living memory.

The shrinkage didn't happen because anyone decided that a smaller navy was better. It happened because ships wear out, and the country stopped building new ones fast enough to replace them.

. . .

A warship isn't like a car. You don't drive it for ten years and trade it in. A properly maintained combatant has a service life of 30 to 40 years. Aircraft carriers last 50 with a mid-life nuclear refueling. Submarines serve until their reactor fuel runs out. These are generational assets. The ships being decommissioned today were commissioned when Reagan or George H.W. Bush was president. The ships being commissioned today will serve until the 2060s.

That means the decisions that determine the size of the fleet in any given decade were made two or three decades earlier. The fleet we have in 2026 was shaped by procurement decisions from the late 1990s and 2000s. And those decisions were shaped by the most dangerous assumption in American strategic history: that history had ended and the good times would roll forever.

After the Berlin Wall fell, the defense budget became the nation's ATM. The "peace dividend" wasn't a strategic concept. It was a bipartisan looting operation. Both parties wanted the money for domestic priorities, and the military couldn't argue against cuts when its primary adversary had just dissolved. Ship procurement dropped from 20-plus hulls a year in the 1980s to single digits in the 1990s. The Navy closed shipyards. Shipbuilding companies merged or went under. The industrial base that had built the 594-ship Navy consolidated down to a handful of surviving yards.

Nobody planned for a future where the fleet would need to grow again. Because the future, according to the best minds of 1995, wasn't going to require one.

. . .

There are exactly four shipyards in the United States that build major combatant vessels. Huntington Ingalls in

Newport News, Virginia, builds aircraft carriers and submarines. General Dynamics' Electric Boat in Groton, Connecticut, builds submarines. Bath Iron Works in Maine builds destroyers. Huntington Ingalls' yard in Pascagoula, Mississippi, builds destroyers and amphibious ships.

That's it. Four yards for the entire Navy.

If you want to understand why the fleet can't grow, start here. It doesn't matter how much money Congress appropriates. You can't build ships without shipyards, and you can't build shipyards without a decade of construction, workforce development, and regulatory certification. The Chinese have been building new shipyards the way Americans build strip malls. We've been arguing about whether to repair the ones we have.

China now has the largest navy in the world by hull count. Over 370 ships and growing. They're not all as capable as American vessels, ship for ship, and the quality gap is real. Chinese submarine crews are far less experienced than American ones, and their boats are noisier. Their carrier aviation program is roughly where the US was in the 1990s. Their joint operations capability has never been tested in combat. The Chinese navy has not fought a naval engagement since 1988. American sailors have been shooting and getting shot at for decades.

Ship for ship, system for system, the US Navy is still the best in the world. But ship for ship doesn't matter when you can't be everywhere and they only need to be in one ocean. The quality advantage is real but depreciating as China gains experience and fields newer platforms. The quantity disadvantage is growing every year and is not reversible on any timeline that matters.

China builds in three years what it takes the US ten to produce. Not because they're smarter. Because they have more yards, more workers, and a government that treats shipbuilding as a strategic priority rather than a jobs program to be fought over in congressional districts.

The workforce problem is the one that makes defense analysts drink. The average age of a skilled shipyard worker in the US is climbing steadily. The apprenticeship programs that used to feed the yards were gutted in the 1990s and never rebuilt. A nuclear-qualified welder takes five to seven years to train. A submarine construction workforce requires security clearances, specialized certifications, and years of on-the-job learning that no classroom can replicate. When these workers retire, the knowledge leaves with them. You can't Google how to weld a submarine hull to nuclear specifications.

The yards are trying to hire. They're offering signing bonuses and relocation packages. But the pipeline of people with the aptitude and willingness to do hard, precise, physically demanding work in a shipyard has been shrinking for a generation. The United States decided in the 1990s that everyone should go to college, and the result is a country with a surplus of communications majors and a critical shortage of the skilled tradespeople who build the things that keep the country safe.

• • •

The most ambitious shipbuilding program currently underway is the Columbia class, the new ballistic missile submarine that will replace the Ohio class boomers that have carried the sea-based nuclear deterrent since the 1980s. The Columbia program is the Navy's number one acquisition priority. It has to be. If the Ohios age out before the Columbias arrive, the United States loses a leg of the nuclear triad. That's not a budget problem. That's an existential one.

The Columbia boats are the largest and most complex submarines ever built by the United States. Each one costs roughly $9 billion. The program calls for twelve boats to replace fourteen Ohios, on the theory that the new design will spend less time in maintenance and more time on patrol. The first Columbia is supposed to enter service in 2031. The program is already over a year behind schedule.

And here's the squeeze: the Columbia boats are being built at the same yard, Electric Boat, that also builds Virginia-class attack submarines. Every month of labor and every square foot of construction space devoted to Columbia is a month and a square foot not devoted to Virginia. The Navy needs both. It's getting less of both than planned because the industrial base doesn't have enough capacity for both programs at once.

What results is a strategic version of robbing Peter to pay Paul. The nuclear deterrent gets built, slowly. The attack submarine fleet shrinks. The attack submarine fleet is what hunts enemy submarines, protects carrier groups, gathers intelligence, and launches Tomahawk strikes. Every one of those missions suffers when the force gets smaller. But nobody is going to delay the nuclear deterrent to build more attack boats. So the attack submarine fleet just keeps shrinking, and the missions keep piling up on fewer hulls.

• • •

The surface fleet has its own version of this problem, and it starts with the ships Congress loves to hate: the Littoral Combat Ships.

The LCS program was supposed to be the affordable alternative to the expensive destroyers. A small, fast, modular ship that could swap mission packages for different tasks. Anti-submarine warfare one month, mine countermeasures the next, surface combat the month after. Cheap to build, cheap to operate, flexible enough to fill multiple roles.

None of that worked out. The ships were plagued with engineering problems. The modular mission packages were delayed, descoped, or cancelled. The two competing designs, built by two different companies in two different states because Congress couldn't pick a winner without angering a delegation, resulted in a fleet of ships that couldn't do their primary missions. The USS St. Louis, a Freedom-class LCS commissioned in 2020, was proposed for decommissioning in 2023 after three years of service. Three years. The USS Sioux City lasted less than five. The USS Little Rock less than six. The USS Detroit less than seven. Ships designed for a 25-year service life were being scrapped before a sailor who joined the crew at commissioning had finished a second enlistment. The Navy proposed decommissioning all nine in-service Freedom-class hulls in 2023, estimating $3.6 billion in savings. Congress blocked some of the retirements, not because the ships were useful, but because decommissioning them meant admitting the whole program was a waste. In the end, the Navy got rid of most of them anyway. The anti-submarine warfare mission package that was supposed to be the Freedom class's reason for existing never worked.

The Navy is now retiring LCS faster than it's being replaced, which means the fleet is shrinking even as new ships come in. The Constellation-class frigates that are supposed to replace the LCS are themselves behind schedule and over budget. The lead ship was originally supposed to deliver in 2026. It won't. The program has experienced the same delays that afflict every major shipbuilding effort: workforce shortages, supply chain disruptions, design changes, and the accumulated rust of an industrial base that hasn't had to perform at scale in decades.

So the Navy is at the same time trying to build nuclear missile submarines, attack submarines, destroyers, frigates, amphibious ships, and support vessels, with a shipyard base that can't deliver any of them on time, using a workforce that's too small and getting smaller.

And into this environment, a war arrives.

• • •

The Iran war didn't add a single ship to the Navy. What it did was accelerate the wear on the ships that already existed while consuming the munitions those ships carry faster than they can be replaced.

The USS Gerald R. Ford had been deployed for over 300 days when it left the war with a fire in its laundry room and mechanical problems that predated the conflict. Three hundred days at sea pushes every system on a carrier toward failure. The crew is exhausted. The machinery is running past its maintenance intervals. The aircraft are flying more hours than their maintenance schedules were designed for. And when the ship finally gets back to port, the repair bill is larger and the repair time is longer because everything that was deferred at sea now has to be done at once.

The Ford isn't alone. Every ship in the strike group was being pushed past its design limits. Destroyers that are supposed to rotate through deployment cycles of seven to nine months were being held in theater longer. Crews that were supposed to get shore time between deployments were being sent back early. John Cordle, a recently retired Navy captain and human factors engineer who specializes in the impact of shipboard fatigue, described a phenomenon that extended-deployment crews call "droning": sailors working almost on autopilot as they cycled through tasks, sleep, and watch standing, their cognitive function degraded by months of cumulative exhaustion. The parent of one Ford crew member told NPR that the crew was told they would be home by early March. Less than twelve hours later, they were told the carrier was being diverted again and probably wouldn't return until May. Some sailors, the parent said, had started questioning whether they wanted to stay in the Navy. The operational tempo that the Iran war demands is grinding the fleet down faster than the shipyards can repair it.

This is the spiral that naval strategists have warned about for years. A fleet that's too small for its commitments gets overworked. Overwork increases maintenance needs. Increased maintenance needs pull ships out of service for longer repairs. Ships in repair aren't available for deployment. The available fleet gets smaller, which means each remaining ship gets worked harder. The spiral accelerates.

The Navy calls this the "maintenance backlog." It's the polite term for a fleet that's eating itself.

. . .

In 2018, Congress passed a law making it national policy to achieve and maintain a 355-ship Navy. It was a nice piece of legislation. Nobody funded it. The Navy later updated its own assessment to 381 ships. The current administration announced something called the "Golden Fleet" goal, the details of which have not been made public. Meanwhile the Congressional Budget Office projected the fleet bottoming out at 285 ships in 2026 and 2027.

Let's be honest about what these numbers mean. They are aspirations unconnected to reality. You don't get from 293 ships to 355 by passing a law. You get there by building roughly 15 ships a year for decades while also maintaining the existing fleet. Current production is nowhere near that rate. The FY2026 budget requested 19 new ships, which sounds close until you realize that the Navy is also decommissioning ships faster than it can commission new ones, and that 19-ship years are the exception, not the norm.

The shipbuilding plan is a wish list presented as a strategy. The strategy requires industrial capacity that doesn't exist, a workforce that hasn't been trained, and a level of sustained bipartisan funding commitment that no Congress has demonstrated in the post-Cold War era. The 355-ship Navy is a bumper sticker. The 293-ship Navy is what we've got.

. . .

Here's what 293 ships means in practice.

The Navy organizes its deployable forces around carrier strike groups and amphibious ready groups. At any given time, roughly a third of the fleet is deployed, a third is in training or working up for deployment, and a third is in maintenance. That's how the rotation works when things are normal.

Things are not normal. The Iran war pulled 40 percent of available carriers into one theater. It pulled destroyers, cruisers, and support ships along with them. It pulled submarines for intelligence gathering and potential strike missions. It pulled amphibious ships for contingency planning. Every asset committed to the Middle East is an asset subtracted from everywhere else.

The Pacific fleet gets thinner. The European presence shrinks. The ships watching the South China Sea, the Taiwan Strait, the Korean Peninsula, and the sea lanes through Southeast Asia are the same ships that are supposed to deter Chinese aggression, reassure allies, and demonstrate American commitment to the region. Pulling them to fight Iran tells China exactly how much American attention they can count on when the next crisis erupts in their neighborhood.

The Marines stationed in Japan who were supposed to be the rapid response force for a Pacific contingency? Two thousand of them redeployed to the Middle East. The THAAD batteries that were protecting South Korea from North Korean missiles? Pulled out and sent to the Gulf. The E-3 AWACS aircraft that provide airborne surveillance across the Pacific? An estimated two-thirds deployed to support the Iran campaign.

This is the real cost of the Iran war. Not just the missiles expended or the money spent. The opportunity cost. Every military asset dedicated to a conflict the American public doesn't support is an asset unavailable for the conflict everyone agrees matters.

. . .

The Navy knows all of this. The admirals aren't stupid. They've been writing papers and testifying before Congress about fleet size for two decades. They've produced force structure assessments, shipbuilding plans, and requirements documents that all say the same thing: the fleet is too small for the strategy.

The response from the political system has been consistent: agree in principle, fund in insufficient quantity, and blame the other party for the shortfall.

Republicans want more ships but also want tax cuts that reduce the revenue available to pay for them. Democrats want ships built in union yards but also want the defense budget redirected to social programs. Both parties agree that the Navy should be bigger. Neither party will make the sustained, multi-decade funding commitment required to make it bigger. Instead they pass one-year budgets, fight over continuing resolutions, and periodically threaten the entire defense enterprise with government shutdowns.

You can't build a navy in one-year increments. A carrier takes seven years from keel laying to commissioning. A submarine takes longer. The workforce needs stable employment to stay in the trade. The supply chain needs predictable orders to maintain capacity. The entire enterprise depends on a level of planning horizon that the American political system is constitutionally incapable of providing.

China doesn't have this problem. When Beijing decides to build ships, the shipyards get orders and the workers show up and the steel gets cut and the hulls go in the water. No congressional delegations fighting over which district gets the contract. No continuing resolutions freezing procurement. No sequestration slashing budgets mid-program. The Chinese Communist Party can plan in decades. The US Congress can barely plan in months.

That's not an argument for authoritarianism. It's an observation about the structural disadvantage a democracy faces in long-term military competition when its political system is broken. The democracy can still win, but only if it gets its act together. And getting its act together starts with acknowledging that 293 ships is not enough, that building more requires sustained commitment, and that the commitment isn't coming from a political system that can't agree on a budget.

• • •

So the Navy is too small, the shipyards can't build fast enough, the workforce doesn't exist, and the political system won't commit to the sustained funding required to fix any of it. The fleet that's supposed to project American power across every ocean is being consumed by one war in one region while the adversary it was designed to deter watches from across the Pacific and takes notes.

But there's one more dimension to this story that makes all the rest of it feel like a warm-up. The ships in that shrinking fleet aren't just expensive to build. They're expensive to lose. And the most expensive of all, the centerpiece of American naval power for eighty years, may be approaching the point where it's too valuable to risk in the very conflicts it was built to fight.

The aircraft carrier. Twelve billion dollars of American prestige floating on a sea full of missiles that cost a thousandth as much.

That's the next chapter. And it's the question the Navy really doesn't want to answer.

Chapter 4: The Carrier Question

My view: build hundreds of frigates, destroyers, and heavy cruisers. Stop chasing the next research project and build the ships that work. Float a couple of battleships for coastal bombardment and psychological effect alone. There is something about a battleship that communicates a kind of seriousness that a drone swarm does not. And we desperately need minesweepers and smaller craft that nobody wants to fund because they are not glamorous. Carriers are good. I am not against carriers. But they take years to build, cost billions, and you only have so many. If one goes down you feel it everywhere. The navy needs mass and variety, not just prestige platforms.

On December 10, 1941, three days after Pearl Harbor, Japanese torpedo bombers found the British battleship HMS Prince of Wales and the battlecruiser HMS Repulse steaming off the coast of Malaya without air cover. Both ships were among the most powerful in the Royal Navy. The Prince of Wales had helped sink the Bismarck six months earlier. They were the pride of British naval power in the Pacific, sent to Singapore to deter Japanese aggression.

The attack lasted about two hours. Both ships sank. 840 men died. It was the first time in history that capital ships operating at sea under full combat readiness were sunk exclusively by air power.

Winston Churchill, who had ordered the ships to Singapore, later wrote that it was the greatest shock he received during the entire war. Not because two ships sank. Because the sinking proved that the weapon around which the Royal Navy had built its entire identity for three centuries, the capital ship with big guns, could be killed by a cheaper, faster, more numerous threat launched from beyond visual range.

The aircraft carrier replaced the battleship within five years. The Yamato, the largest battleship ever built, was sunk by aircraft on its final mission without ever firing its main guns at an enemy ship. By 1945, everyone understood that the era of the gun battleship was over. The carrier was the new capital ship.

Eighty years later, the question is whether the carrier is about to follow the battleship into history. And the Iran war is the stress test that's supposed to answer it.

• • •

The answer, so far, is complicated. And the complication is dangerous because it lets people hear what they want to hear.

The carriers survived. No Iranian missile hit a US carrier. Iran launched over 500 ballistic missiles in the first five days and none of them found a carrier strike group. The IRGC claimed they hit the USS Abraham Lincoln. US Central Command called it a lie. Based on the available evidence, the carriers were never seriously threatened by direct attack.

Carrier advocates will use this to argue that the platform is vindicated. See? We sent two carriers into a shooting war and both came home. The defense works. The Aegis escorts do their job. The electronic warfare bubble is impenetrable. The carrier is still king.

That argument has a problem the size of the Persian Gulf in the middle of it.

Iran couldn't hit the carriers because Iran couldn't find the carriers. Its military lacked the one thing you need to kill a ship at long range: persistent over-the-horizon targeting. You can have all the anti-ship missiles in the world, but if you don't know where the ship is, you're shooting into empty ocean. Iran didn't have the satellite constellation, the maritime surveillance aircraft, the ocean-scanning radar, or the reconnaissance drone network needed to maintain a continuous targeting solution on a carrier strike group operating hundreds of miles offshore in an electronic warfare environment designed specifically to make detection impossible.

Iran was firing blind. The carriers survived not because carriers are invulnerable, but because this particular adversary couldn't complete the kill chain.

China can.

. . .

The People's Liberation Army has spent twenty years and untold billions building a military system whose primary purpose is killing American aircraft carriers. They don't talk about it in those terms. They call it "anti-access/area denial," or A2/AD. But the meaning is simple: if a US carrier strike group tries to operate within range of China's coast, China intends to find it, track it, target it, and sink it.

The tools they've built for this are formidable. The DF-21D, which Western analysts call the "carrier killer," is a medium-range ballistic missile with a maneuverable warhead designed to hit a moving ship at sea. The DF-26 extends that capability to longer range, potentially covering the entire Western Pacific from launchers deep inside China. These weapons fly on ballistic trajectories at hypersonic speeds, making them extremely difficult to intercept. The warhead maneuvers during its terminal phase, adjusting its aim as the target moves.

But the missiles are just the sharp end. Behind them sits the targeting architecture that Iran lacked. China has a constellation of surveillance satellites, including synthetic aperture radar birds that can image the ocean surface through clouds and darkness. It has over-the-horizon radar systems that can detect large surface vessels at enormous range. It operates long-range maritime patrol aircraft and reconnaissance drones. It has submarines that can trail a carrier strike group and relay position data. And it has the command and control systems to fuse all of this sensor data into a real-time targeting picture and pass it to the shooters.

China can complete the kill chain. Whether it can do so reliably, against the full electronic warfare suite of a carrier strike group with Aegis escorts, in the chaos of actual combat with countermeasures and jamming and decoys, is genuinely unknown. Nobody has ever tried. The Chinese haven't tested the system end-to-end against a real, defended carrier because you can't do that without starting a war. And the Americans can't be sure their defenses will work because they've never been tested against the full weight of a Chinese anti-ship barrage.

But the uncertainty cuts both ways. The US can't be sure the carrier survives. And that uncertainty, by itself, changes the strategic calculus.

. . .

Think about what a carrier costs. Not just money, though the money is staggering. The USS Gerald R. Ford cost $13.3 billion to build. Its air wing costs several billion more. The escort group that travels with it, four or five destroyers and cruisers, adds tens of billions. The supply ships, the submarine escort, the supporting infrastructure ashore. A carrier strike group represents perhaps $30 to $40 billion in total assets afloat.

Now think about the people. A carrier has a crew of over 5,000. The air wing adds hundreds more. The escorts carry another thousand or so each. A strike group puts roughly

7,000 to 8,000 American service members in one concentrated location.

If a carrier sinks, you lose a ship that takes eight years to replace, assuming the shipyard capacity exists. You lose an air wing of 70-plus aircraft. You potentially lose thousands of sailors and aviators, the largest single loss of American military personnel in combat since World War II. And you can't replace any of it on a timeline that matters for the war you're fighting.

The strategic shock would be seismic. Aircraft carriers have been the symbol of American global power projection since 1945. No American carrier has been sunk in combat since the Hornet went down at Santa Cruz in 1942. The psychological impact of losing one, in the age of live satellite imagery and social media, would be beyond anything the American public has experienced. It would be the military equivalent of 9/11 at sea.

And every adversary's risk calculus would reset overnight. The Chinese generals who built the DF-21D would see their investment validated. The proliferation of carrier-killing technology would accelerate. Iran was already in talks with China for CM-302 anti-ship cruise missiles before the war. After a carrier sinking, every coastal nation with a grievance would be shopping for the weapon that brought down Goliath.

• • •

The Ford didn't get sunk. It got something almost as instructive: it broke down.

The fire in the laundry room on March 12 was the visible failure. The sewage system that had been malfunctioning before the ship even arrived in theater was the less dramatic but equally telling one. An NPR investigation obtained internal Navy documents through a Freedom of Information

Act request showing that "every day that the entire crew is present on the ship, a trouble call has been made" to repair or unclog the vacuum sewage system, going back to June 2023. Vice Chief of Naval Operations Admiral Jim Kilby told the Senate Armed Services Committee in March 2026 that the deployment would ultimately stretch to roughly 11 months, approaching records set during the Vietnam War. The Ford is the most technologically ambitious warship in history, bristling with systems that had never been tested at scale in a real deployment: an electromagnetic aircraft launch system replacing steam catapults, an advanced weapons elevator system, a new-design nuclear reactor, a dual-band radar, and a level of automation that cut the crew by several hundred compared to the Nimitz class.

Many of these systems didn't work as advertised. The electromagnetic catapult had reliability issues. The weapons elevators, which move bombs and missiles from the magazines to the flight deck, were years behind schedule and kept failing. The Ford spent much of its early career going back to the shipyard for fixes rather than deploying. When it finally deployed for the Iran war, it had been in service for nine years and still wasn't fully operational.

Thirteen billion dollars. Nine years. Not fully operational.

The Ford withdrew from combat operations around March 17 and sailed to Crete for repairs. The most expensive warship ever built, in its first real war, sidelined by mechanical failures. Not by an enemy missile. By its own laundry and plumbing.

The Navy sent the George H.W. Bush as a potential replacement. It extended the Nimitz past its planned decommissioning. It scrambled to maintain carrier presence in a theater that was devouring every asset sent to it. The system worked, barely, through improvisation and overextension. But it worked only because Iran couldn't exploit the vulnerability. A more capable adversary watching a carrier limp out of a war zone would see something very different: an opportunity.

• • •

Here's the question that the carrier's defenders don't want to confront: if the carrier is too expensive to lose, and the adversary has built weapons specifically designed to sink it, at what point does the carrier become too valuable to use?

A weapon you can't risk in combat is not a weapon. It's a liability. A $13 billion ship that has to stay 500 miles from the enemy coast to avoid the DF-21D threat envelope can still launch aircraft, but those aircraft now need mid-air refueling to reach their targets, which reduces sortie rates and increases complexity. A carrier that has to dedicate half its escorts to missile defense rather than offensive operations has its combat power halved. A carrier whose loss would be so strategically devastating that commanders are reluctant to put it in harm's way has already been deterred.

This is what the Chinese call "the assassin's mace." You don't have to sink the carrier. You just have to make the Americans believe you can. The threat changes behavior even if the weapon is never fired. The DF-21D has never been used against a ship. It doesn't need to be. Its existence pushes American carriers further from the Chinese coast, reduces their effectiveness, and forces the Navy to allocate enormous resources to defending them rather than projecting power.

The carrier is becoming a hostage to the fleet it's supposed to command.

• • •

The historical parallel to the battleship is irresistible, and the Navy hates it. Because the battleship admirals of 1941 also believed that their platform was indispensable. They also had rational arguments for its continued relevance. The battleship could absorb punishment that no other vessel could survive. Its guns could devastate shore targets. Its very presence commanded respect and projected power. All of those arguments were true, and all of them were irrelevant the moment aircraft could find and kill a battleship from beyond gun range.

The carrier could follow the same trajectory. Not in a day, not with one dramatic sinking, but through a gradual shift in the cost-benefit calculus that makes it progressively less rational to build $13 billion targets in a world filling up with million-dollar missiles designed to kill them. The aircraft carrier might remain the most powerful single weapon system afloat for another twenty years. The question is whether it remains the most cost-effective way to project naval power. And the answer to that question is increasingly uncertain.

The optimistic case isn't that the carrier will live forever. It's that the Navy recognizes the shift and adapts before a catastrophe forces the issue. Distributed lethality: instead of concentrating combat power in a few enormous, vulnerable platforms, spread it across many smaller, cheaper, more expendable ones. Unmanned systems: robotic aircraft and surface vessels that can absorb the risk that carriers increasingly can't. Directed energy: lasers and electronic warfare systems that change the cost equation for missile defense. Subsurface platforms: submarines that are inherently harder to find and target than surface ships.

And there's a fifth option that nobody in the Navy talks about because it isn't sexy. The armored surface combatant. Call it what it is: a battleship. Not the World War II kind with sixteen-inch guns, though the shore bombardment gap is real enough that the gun question deserves revisiting. The modern kind: a heavily armored hull built to absorb hits and keep fighting rather than to avoid detection and pray nothing gets through. The Navy spent $23 billion on three Zumwalt destroyers whose guns don't have working ammunition. The Iowa-class battleships that those Zumwalts theoretically replaced took kamikaze hits and kept steaming. The Ford got knocked out by a laundry fire. An Iowa has twelve inches of belt armor. A Burke has none. It's designed not to get hit. When the don't-get-hit strategy fails, you lose the ship.

A modern armored combatant carrying railguns, directed energy weapons, VLS cells, and enough belt armor to survive multiple missile hits would address the fragility problem that the entire precision-over-mass doctrine has created. It would provide the shore bombardment capability the Marines have been begging for since the Iowas retired. And it would put something in the water that scares people. A carrier over the horizon is abstract. A battleship parked offshore where you can see it through binoculars, with barrels pointed at your coastline, communicates something that a stealth destroyer never will. Reagan understood that when he recommissioned all four Iowas. Deterrence works better when it's visible.

Nobody is pursuing this because the carrier admirals control the budget and battleships compete for the same funding. The surface warfare community wants more Burkes. The submarine community wants more Virginias. The carrier community wants more Fords. Nobody is arguing for a platform that doesn't advance their particular community's institutional interests. The battleship has no constituency because it has no admiral, and it has no admiral because the Navy eliminated the surface gunnery career path forty years ago. The most logical platform for a world of cheap missiles and expensive fragile ships is the one platform the Navy's institutional culture won't consider.

All of these alternatives exist in various stages of development. Some are being deployed in limited numbers. None are at the scale needed to replace carrier-based aviation as the primary instrument of American naval power projection. Because replacing the carrier means replacing the Navy's identity, its institutional culture, its promotion pathways, its budget allocation model, and its congressional patronage network. That's not a technology problem. That's a political and cultural problem, and those are harder to solve than engineering challenges.

The battleship admirals never willingly gave up their battleships. The carrier admirals had to drag the platform into being against institutional resistance. The cycle may be repeating: the future of naval warfare is probably unmanned, distributed, and submarine-heavy, and the carrier admirals will resist it the same way the battleship admirals resisted them.

· · ·

This is where Part One closes, and it closes with a paradox.

The US military is the most technologically advanced fighting force in human history. Its weapons are more precise, more lethal, and more sophisticated than anything any other country fields. Ship for ship, plane for plane, system for system, it can beat any opponent on the planet.

And it's running out of missiles in a war against a country whose entire military budget is a rounding error on the Pentagon's spreadsheet. Its fleet is the smallest it's been in a century and shrinking during the war. Its newest carrier got knocked out by a laundry fire. Its production base can't surge. Its shipyards can't build. Its workforce doesn't exist. And the adversary it was designed to fight hasn't even entered the arena.

The paradox is this: American military power is at the same time the strongest and the most fragile it has ever been. Strongest in capability. Most fragile in capacity. It can do anything, for about three weeks, and then the math takes over.

None of it is unfixable. Lasers that kill drones at the cost of diesel fuel. Autonomous swarms that flip the cost-exchange ratio. Production lines that can surge if the investment is sustained. The engineering exists. The designs exist. What doesn't exist is the institutional will to deploy them fast enough. That will is the subject of the rest of this book.

How did the most powerful country on Earth get into this position? The answer isn't military. It's political. It's about why we fight the wars we fight, who benefits from them, and what we've given up to wage them.

That's Part Two.

Part Two: Wars Without Reasons

Chapter 5: What Are We Doing Here?

My honest answer to the question of what we are doing out there: we are stagnated. Stuck in the same ruts we have been stuck in for thirty years, driven by defense contractors and political interests that have nothing to do with the national interest. We are energy independent now. We have been for years. So what exactly are we doing in the Middle East? That is the question nobody in Washington will answer plainly, because the honest answer is that we are there for reasons that made sense in 1990 and have not been seriously re-examined since. We are also far too controlled by our economic relationship with China to make the hard strategic calls that relationship requires. The stagnation is not accidental. It serves people who profit from it.

On March 9, 2026, ten days into the war with Iran, the Quinnipiac University poll asked American voters a simple question: do you support or oppose the US military action against Iran?

Fifty-three percent opposed. Forty percent supported. The rest weren't sure.

That's not how wars are supposed to work in a democracy. When a president sends Americans into combat, the expectation, the historical norm, is a surge of public support. Rally around the flag. Back the troops. Ask questions later.

Pearl Harbor gave Roosevelt near-unanimous support. 9/11 gave George W. Bush approval ratings in the nineties. Even the 2003 invasion of Iraq, launched on evidence that turned out to be fabricated, initially had majority support.

The Iran war never got its rally. From the first day, more Americans opposed it than supported it. And the numbers got worse as the war went on.

Why? Because nobody could explain what we were doing there.

• • •

The Washington Post asked a version of that question twice in the first two weeks: has the Trump administration clearly explained the goals of US military action in Iran?

Sixty-five percent said no. Both times. The number barely moved between the two surveys, which means the administration wasn't even gaining ground on the explanation front as the war progressed. Two-thirds of the country, after two weeks of combat, still didn't understand the point.

That's not a communication failure. When you can't explain your war aims after two weeks, the problem isn't messaging. The problem is that the aims don't hold together.

Listen to the justifications as they rolled out, and you can hear them shifting like a kaleidoscope. It was about Iran's nuclear program. Except the US and Israel had already bombed nuclear facilities during the twelve-day war in June 2025. It was about supporting the Iranian protesters who had been killed by their government. Except the strikes began weeks after the protests had been suppressed and nobody had proposed military action while the protesters were dying. It was about an imminent military threat to the United States. Except when Quinnipiac asked whether Iran posed an imminent threat, 55 percent of voters said no.

The stated reason kept changing because no single reason was strong enough to bear the weight of what was happening. A full-scale air and naval campaign, the largest American military deployment to the Middle East since the 2003 invasion of Iraq, carrier strike groups, B-2 bombers, hundreds of cruise missiles, thousands of service members, a war that closed the Strait of Hormuz and sent oil prices through the roof. That's not a proportional response to any of the individual justifications offered. It's a war in search of a reason.

● ● ●

The polling tells a deeper story than just opposition to this specific war. It reveals something structural about what Americans believe military force is for and when it's worth the cost.

Multiple polls asked versions of the same question: will this war make America safer or less safe? The answers were devastating for the administration. CNN found 54 percent thought the strikes would make Iran more of a threat, not less. Quinnipiac found 47 percent said the war made the US less safe, versus 34 percent who said safer. Even the Washington Post poll, which was the most favorable to the administration on other questions, found 53 percent saying the war would not contribute to long-term US security.

Americans weren't just opposed to the war. They thought it was making things worse. That's a different category of opposition. You can be against a war because you think it costs too much or takes too long while still believing it serves a purpose. When the public thinks the war is actively counterproductive, you've lost the argument at a level that no amount of messaging can fix.

And the casualties hadn't even mounted yet. At the time of the early polls, the Pentagon had reported seven service members killed and about 140 wounded. By the standards of American wars, those numbers are small. But when the Post asked whether the casualties were acceptable given the goals and costs of the war, more than six in ten Americans said unacceptable. Americans were less tolerant of casualties in the first two weeks of the Iran war than they had been during the early phase of the Iraq War, a conflict that also turned sour but at least started with majority support.

What differs is purpose. People will accept sacrifice for a cause they understand and believe in. They will not accept sacrifice for a cause nobody can explain.

The domestic political damage compounds with every week the war continues. Gas prices are the most visible transmission mechanism. When the Strait of Hormuz closed and oil spiked past $120, the average American didn't think about geopolitics or nonproliferation or the balance of power in the Persian Gulf. They thought about the number on the pump. And that number, which the administration cannot control because it is set by global markets, becomes the daily referendum on whether the war is worth it. Every dollar added at the pump is a vote against the war, cast not in a ballot box but in a gas station, and it hits hardest in the rural and working-class communities that the administration depends on for its political base.

The partisan split on the war is revealing. In most American wars, support tracks with the president's party: Republicans supported Iraq under Bush, Democrats supported intervention in Libya under Obama. The Iran war fractured even that pattern. Republican voters were divided, with the populist wing that powered Trump's rise openly skeptical of another Middle Eastern war and the hawkish establishment wing supporting it. Democratic opposition was near-universal, not because Democrats had a coherent alternative policy but because opposing the war was the path of least political resistance. The result was a war without a domestic political constituency. Hawks supported the objectives but worried about the cost. Populists opposed foreign entanglement on principle. The left opposed it on humanitarian grounds. The center checked the gas price and the poll numbers and concluded they were against it too. The only Americans who supported the war with anything resembling enthusiasm were a narrow slice of national security professionals and the evangelical communities whose theological commitment to Israel made the alliance non-negotiable.

A war without domestic support is a war on a timer. The administration knows this. The military knows this. The adversary knows this. Iran's strategy, to the extent it has one beyond survival, is to make the war expensive enough and long enough that American domestic politics forces a withdrawal. It does not need to defeat the American military. It needs to outlast American patience. And American patience, as measured by every available poll, was already running out before the first month was over.

Foreign information operations accelerate the timer. Disinformation campaigns do not need to create division from scratch. They are force multipliers that amplify fractures that already exist. A country where 53 percent oppose a war on its first day is a country where the fractures are already wide open. Iranian, Russian, and Chinese state-affiliated networks can pour content into those cracks through social media, targeting the populist right with anti-intervention messaging, targeting the left with civilian casualty imagery, targeting veterans with waste-and-corruption narratives, and targeting everyone with gas price outrage. None of these narratives need to be fabricated from nothing. They just need to be selected, amplified, and pushed into the algorithmic feeds of audiences already primed to receive them.

The campaigns work not because Americans are gullible but because the information environment is designed to reward engagement over accuracy, and outrage engages more than nuance. The loop closes when domestic politicians adopt the foreign-amplified narratives for their own purposes, repeating talking points that originated in a troll farm because the talking points poll well with their base. The polarization deepens. The next campaign finds even wider cracks to exploit.

A divided country fighting an unpopular war in a media ecosystem optimized for polarization is the ideal target for information warfare. The United States in March 2026 checks every box.

• • •

The pattern didn't start with Iran. It started with Iraq.

The 2003 invasion had a clear stated objective: eliminate weapons of mass destruction that posed an imminent threat to the United States. The objective turned out to be based on intelligence that was wrong, manipulated, or both. The weapons didn't exist. But the war continued for eight years, morphing from WMD elimination to regime change to nation building to counterinsurgency to an open-ended commitment that nobody could define or justify but nobody could figure out how to end.

Afghanistan was worse. The original objective, destroying al-Qaeda's base of operations and killing the people who planned 9/11, was achieved within months. Everything after that was mission creep on a civilizational scale. Building a democratic government in a tribal society. Training a national army that collapsed the moment American support was withdrawn. Twenty years, two trillion dollars, and 2,400 American dead for a country that reverted to Taliban rule within weeks of the last helicopter leaving Kabul.

Libya. Obama authorized air strikes to prevent a massacre in Benghazi. The mission expanded to regime change. Gaddafi was killed. The country descended into civil war and became a failed state. Nobody had a plan for what came after because nobody had planned to do what they ended up doing.

Syria. American troops deployed to fight ISIS, then stayed to counter Iranian influence, then stayed because nobody could articulate a reason to leave or a condition under which they would leave. They're still there. Ask a Pentagon spokesperson what the mission is and you'll get an answer that could mean anything.

The Iran war fits this pattern so precisely it's almost formulaic. Dramatic opening strike. Vague and shifting objectives. No defined end state. No exit criteria. No answer to the question that every war plan is supposed to start with: what does success look like, and how will we know when we've achieved it?

. . .

There's a reason this keeps happening, and it's not stupidity. The people who plan and launch these wars are not dumb. They're operating within a system that incentivizes action and penalizes clarity.

A president who defines clear war aims can be held accountable for failing to achieve them. A president who keeps the objectives vague can declare victory at any point and blame someone else if things go wrong. Bush couldn't find WMDs, so the war became about democracy. When democracy didn't take hold, the war became about stability. When stability proved elusive, the war just kept going. Each redefinition bought time and shifted the goalposts.

The defense establishment benefits from ambiguity because ambiguity sustains budgets. A war with a clear end state has a clear end date, and an end date means budget cuts. A war with vague objectives and no exit criteria can last indefinitely, which means indefinite procurement, indefinite deployment, and indefinite revenue for the contractors who supply the machine. Nobody in the military-industrial complex has an incentive to end a war. They have every incentive to prolong it.

Congress benefits from ambiguity because it avoids accountability. The War Powers Act requires congressional authorization for sustained military action, but every president since its passage has found ways to circumvent it, and Congress has let them. Voting for a war means owning the consequences. Not voting means you can criticize the president whether it goes well or badly. The path of least resistance is to let the executive wage war without authorization and then hold hearings about it afterward.

And the media benefits from ambiguity because war is good for ratings. The breathless coverage, the retired generals on panels, the maps and graphics and correspondent packages from aircraft carriers. A war with a clear purpose and a quick resolution is a brief ratings bump. A war that drags on with shifting narratives is a content engine that runs for months or years.

Every institution in the system has an incentive to start wars and no incentive to define them clearly enough to end them. What results is a twenty-five-year pattern of military actions that begin with a bang, continue without purpose, and end with a shrug.

• • •

The human cost of this pattern falls on people who don't benefit from it.

The service members who deployed to the Middle East for the Iran war are the same people, or the same types of people, who deployed for Iraq and Afghanistan. Professional volunteers from military families, small towns, and working-class communities. They go where they're sent, do what they're told, and trust that the people sending them have a reason that justifies the risk. When the reason turns out to be incoherent, they bear the cost anyway.

Thirteen American service members killed as of mid-March, with the wounded count climbing past the 140 reported in the war's first two weeks. A refueling aircraft crashed in western Iraq, killing all six crew members. Iranian missiles targeted the headquarters of the US Navy's Fifth Fleet in Bahrain multiple times. These aren't statistics. They're people who were sent to fight a war that most of their fellow citizens opposed and couldn't explain.

The economic cost falls on everyone. Oil prices spiked. Gas approached four dollars a gallon. The Strait of Hormuz closure disrupted global supply chains. Fertilizer prices surged during planting season. Inflation, which was already eroding living standards, got another accelerant. The Americans paying more at the pump and more at the grocery store are subsidizing a war they didn't ask for and don't understand, waged on behalf of objectives nobody has clearly defined.

YouGov found that 61 percent of Americans wanted to prioritize ending the war as quickly as possible, versus only 24 percent who wanted to continue fighting until all objectives were achieved. The public wasn't divided on this. A supermajority wanted out. The question is why they weren't getting out.

• • •

The answer involves the relationship between the United States and its closest ally in the region. Because while America's war aims were vague and shifting, someone else's war aims were crystal clear. Someone who understood exactly what they wanted from this conflict, exactly how to get it, and exactly how to keep the United States committed long past the point where American interests were served.

Ninety-three percent of Israeli citizens supported the war in Iran. They knew what they were fighting for. The question is whether what they were fighting for had anything to do with what Americans thought they were fighting for.

It didn't. And that's the problem.

Chapter 6: The Tail Wags the Dog

In early 1941, Britain was losing the war. The Luftwaffe had failed to break the RAF in the Battle of Britain, but that was a defensive victory, not a path to winning. The British Army had been thrown off the continent at Dunkirk. The U-boats were strangling supply lines across the Atlantic. The treasury was nearly empty. The empire was overstretched. Without the United States, Britain would eventually negotiate some form of accommodation with Nazi Germany or simply be ground down into irrelevance.

Winston Churchill understood all of this with perfect clarity. He also understood that the United States, under Franklin Roosevelt, had no political appetite for entering a European war. The American public was isolationist. Congress was hostile to intervention. Roosevelt himself was sympathetic to Britain but constrained by an electorate that had no interest in dying for someone else's fight.

So Churchill set about the most consequential manipulation of a larger ally in modern history. He didn't beg. He didn't plead. He did something more effective. He figured out what Roosevelt needed emotionally and psychologically, and he packaged Britain's war as the fulfillment of that need.

Roosevelt needed to feel like a historic figure defending civilization itself. Churchill gave him that narrative. Roosevelt needed to believe that American intervention was his idea, not a British request. Churchill arranged events so that each escalation felt like an American choice. Lend-Lease. Convoy escorts that were effectively acts of war. The Atlantic Charter. Each step made the next one harder to refuse, each one framed not as helping Britain but as defending American values. By December 1941, the US was so entangled that Pearl Harbor didn't drag America into the war so much as provide the final justification for a commitment that was already functionally made.

Churchill was brilliant. He saved his country. And the playbook he wrote is being run again, eighty-five years later, by a man who studied it carefully.

. . .

Benjamin Netanyahu has been warning about Iran for thirty years. He brought poster boards to the United Nations. He drew cartoon bomb diagrams. He told every American president from Clinton onward that Iran was months away from a nuclear weapon and that only decisive action could stop it. Some of those presidents listened politely. Some ignored him. None of them gave him what he wanted: an American military commitment to destroy Iran as a regional power.

Then Donald Trump came along.

Netanyahu understood Trump the way Churchill understood Roosevelt. Different man, different levers, same operation. Roosevelt responded to moral framing and historical legacy. Trump responds to flattery and the desire to be seen as the strongest, toughest, most decisive leader in the room. Churchill gave Roosevelt the narrative of civilization's defender. Netanyahu gives Trump the narrative of the warrior president who finished what everyone else was too weak to start.

The flattery campaign has been running for years. Netanyahu named a settlement after Trump. He put up billboards of the two of them together during Israeli elections. He was the first foreign leader to praise Trump's "strength" and "courage" after every decision. The Abraham Accords, which Netanyahu helped broker, were packaged to ensure

Trump got maximum credit. Trump still talks about them as his greatest foreign policy achievement.

The relationship is built on a single insight: Trump doesn't care about policy details. He cares about being told he's great. Netanyahu provides that service more skillfully than anyone else on the world stage, and the price of the service is American military power deployed in the service of Israeli strategic objectives.

. . .

The divergence between American and Israeli war aims in the Iran conflict is not subtle. It's a chasm.

The American position, to the extent one was ever articulated, was about denuclearization and deterrence. Defense Secretary Pete Hegseth said on March 2 that "this is not a so-called regime change war." Trump gave conflicting signals, sometimes talking about regime change and sometimes about deals, but the Pentagon's stated objectives were limited: degrade Iran's military capability, destroy its nuclear infrastructure, and create conditions for negotiation.

The Israeli position was regime change. Period. Netanyahu described the operation as an effort to destroy Iran as a regional power permanently. He talked about "breaking their bones." He addressed the Iranian people in Farsi, calling on them to "come to the streets in your millions to overthrow the regime of fear." An Israel Democracy Institute poll found 93 percent of Jewish Israelis supported the war, with most believing it should continue until the regime was overthrown.

Ninety-three percent. In a country where people can't agree on what day of the week it is, 93 percent agreed on regime change in Iran. That's not public opinion. That's national consensus, and it's a consensus that extends far beyond Netanyahu's political base. Even the Israeli opposition was on board. Opposition leader Yair Lapid visited a missile impact site and said that eliminating all of Iran's ballistic capabilities was a defined objective "and we cannot stop until that happens." When the opposition is telling you to keep bombing, there's no domestic brake.

The Israeli military told NPR it needed "several more weeks" to complete its war aims and described current progress as "halfway there." Halfway to what? Not halfway to the limited American objectives, which were largely achieved in the first week. Halfway to the destruction of Iran as a functioning military power, which is an Israeli objective the United States never endorsed.

. . .

Netanyahu went further than just defining different objectives. He actively expanded the war in ways the United States hadn't sanctioned.

Israel struck Iran's South Pars gas field unilaterally. This wasn't a military target in the traditional sense. It was energy infrastructure, an escalation designed to put economic pressure on Iran that went beyond what the US campaign was intended to do. When asked about it, Netanyahu said "President Trump asked us to hold off on future attacks, and we're holding out." The phrasing was revealing. He acknowledged that Trump had told him to stop, then implied he was graciously complying, while having already done the thing Trump asked him not to do.

Israel expanded the war into Lebanon, hitting Hezbollah targets and pushing ground forces toward a "Rafah model" scorched-earth policy in the south, reportedly aimed at depopulating a buffer zone covering 10 percent of Lebanon's landmass. Israeli officials were pursuing objectives in Lebanon that had nothing to do with the Iran war and everything to do with long-standing Israeli security goals that predated the conflict.

Netanyahu even floated the possibility of a ground component. "It is often said that you can't win, you can't do revolutions from the air," he told reporters. "That is true. You can't do it only from the air." He was signaling willingness to put troops on Iranian soil, something the United States had explicitly said it would not do, and something that would transform the conflict from an air campaign into an open-ended occupation.

Each escalation served the same purpose: make it harder for the United States to disengage. Every time Israel expanded the war's scope, the status quo shifted. Every Iranian retaliation against Israeli escalation hit American bases and allies, which demanded an American response, which deepened American involvement, which made withdrawal more politically costly. The ratchet only turned one direction.

• • •

The Churchill parallel extends to the personal dynamics. Churchill could be charming, solicitous, and deferential to Roosevelt in person while ruthlessly pursuing British interests behind the scenes. He flattered Roosevelt's ego while manipulating the strategic environment to foreclose alternatives. He made sure that at every decision point, the path of least resistance for Roosevelt was deeper commitment.

Netanyahu operates the same way with Trump. Every public statement from Jerusalem frames the war as Trump's historic achievement. Trump destroyed Iran. Trump is the strongest president in history. Trump is reshaping the Middle East. Netanyahu is ghostwriting Trump's legacy narrative in real time, and the price of admission is continued American military support for objectives Trump never clearly endorsed.

It works because Trump is uniquely susceptible. Most presidents have policy frameworks, institutional advisors, and strategic worldviews that create some resistance to personal manipulation. Trump has none of these. His decision-making runs on personal relationships, ego gratification, and whatever he saw on television most recently. A foreign leader who provides constant flattery, public praise, and a narrative of strength and victory has more influence over Trump's decisions than his own national security establishment.

What results is a war whose scope, duration, and objectives are being defined by an ally whose interests diverge from America's, enabled by a personal relationship between two leaders where one is playing the other like a fiddle.

• • •

The irony is that the United States has all the leverage and none of the will to use it.

Israel cannot sustain this campaign alone. Its air force has range limitations without American tanker aircraft to refuel them over Iranian airspace. Its Iron Dome and David's Sling interceptors depend on American-funded and American-manufactured components. The 1,200 bombs Israel dropped in the first 24 hours came from American stockpiles. If the United States said "we're done, and the resupply pipeline closes in 30 days," the Israeli air campaign would be over within weeks.

Any American president could end this war by picking up the phone. The question is why this one won't.

The surface answer is domestic politics. AIPAC and the broader pro-Israel lobby infrastructure is arguably the most effective single-issue political operation in the country. Any president who publicly cuts Israel off mid-war faces coordinated, massively funded political assault.

Congressional resolutions. Primary challenges. Wall-to-wall media framing of "abandoning our ally." The evangelical base, a huge chunk of Trump's coalition, views Israel through an eschatological lens that makes support non-negotiable.

The deeper answer is the one this chapter is about. Netanyahu has constructed a psychological relationship with Trump that makes disengagement feel like personal betrayal rather than strategic recalibration. Trump doesn't evaluate the Israel relationship through a policy lens. He evaluates it through a personal one. Netanyahu is nice to me. Netanyahu says I'm strong. Netanyahu gives me credit. Cutting off Netanyahu means losing someone who tells me what I want to hear. That's not how a president should make decisions about war and peace. It's how this president makes every decision about everything.

• • •

There is one mechanism that might break the dynamic, and it's not strategic wisdom or public pressure or congressional action. It's Trump's ego getting bruised.

Trump has turned on loyal allies before. Jeff Sessions. Bill Barr. Mike Pence. Mitch McConnell. Fox News. People who were intensely loyal until the moment Trump felt they weren't sufficiently loyal to him. The betrayal didn't have to be real. It just had to feel real. And once Trump turns, he doesn't turn back.

If Netanyahu does something that makes Trump feel disrespected, upstaged, or insufficiently grateful, the entire dynamic inverts overnight. The South Pars gas field strike was a potential crack. Netanyahu hit a target Trump told him not to hit. Trump let it slide. But if Iran retaliates against the Gulf states in a way that spikes oil to $120 a barrel and craters Trump's approval further, Trump might suddenly remember that Netanyahu went rogue, and the flattery won't be enough.

That's the off-ramp. Not policy. Not strategy. Not public opinion. A narcissist's wounded ego. It's a terrifying thing to depend on for ending a war. But in the structural reality of this presidency, it might be the only thing that works.

. . .

The Churchill comparison has one more dimension that matters for where this story goes.

Churchill's manipulation of Roosevelt served a cause that most of the world ultimately agreed was just. Defeating Nazism justified the methods. History largely forgave Churchill because the alternative, a Nazi-dominated Europe, was genuinely worse.

But Churchill's success came with a cost he didn't plan for. The war saved Britain from conquest but destroyed it as a global power. The empire dissolved within twenty years. The financial cost of the war left Britain dependent on American aid for a generation. Churchill won the battle and lost the empire.

Netanyahu may be running the same trade. He's getting what he's wanted for thirty years: the United States destroying Iran as a regional power. But the cost is being paid in American munitions, American readiness, American credibility, and American attention. Every SM-6 fired over the Persian Gulf is an SM-6 that won't be available in the Pacific. Every carrier deployed to the Middle East is a carrier not deterring China. Every dollar spent on this war is a dollar not spent rebuilding the industrial base or the fleet.

If you burn out your patron's military and political will fighting your war, what happens when the next threat emerges and the patron doesn't show up?

Netanyahu is too good at getting what he wants. And what he wants may be the thing that destroys what he needs.

That's the risk of being the tail that wags the dog. You get to set the direction. But you're still attached to the dog. And if the dog collapses from exhaustion, you go down with it.

Chapter 7: The Lesson Every Country Learns

In 2003, Muammar Gaddafi did something that Western diplomats called courageous and far-sighted. On December 19 of that year, Libya's Foreign Ministry announced that the country would voluntarily dismantle its nuclear weapons program, its chemical weapons stockpile of 25 metric tons of mustard gas, and its longer-range ballistic missiles. The decision followed months of secret negotiations with the United States and Britain, catalyzed partly by the October 2003 interception of a ship carrying centrifuge components from the A.Q. Khan nuclear black market network destined for Tripoli. In exchange, sanctions were lifted. Diplomatic ties were restored. Libya rejoined the international community. Gaddafi was praised as a model of what a rogue state could become if it chose cooperation over confrontation. Paula DeSutter, the State Department's verification chief, told Congress the administration wanted Libya to be "a model for other countries."

Eight years later, on March 19, 2011, NATO began bombing his country under UN Security Council Resolution 1973, authorized to protect civilians. The mission expanded to regime change. On October 20, 2011, French fighter jets destroyed his convoy as he fled Sirte. He was dragged from a drainage pipe by rebels, beaten, sodomized with a bayonet, and shot dead. The video circulated worldwide. Secretary of State Hillary Clinton watched the footage and said, with a laugh, "We came, we saw, he died."

Libya has been a failed state ever since. Three competing governments, an ongoing civil war, open-air slave markets, and a humanitarian catastrophe that no one in Washington talks about because there's no political upside in reminding people what happens after we break things.

Every government on Earth took note. Not of the humanitarian disaster. Of the sequence. Give up your nuclear program. Get praised. Get bombed. Get killed.

• • •

In 1994, Ukraine agreed to the Budapest Memorandum. At the time, it held roughly 1,900 nuclear warheads inherited from the Soviet Union, making it the third-largest nuclear arsenal in the world. In exchange for giving up those weapons, Ukraine received security assurances from Russia, the United States, and the United Kingdom. The three powers committed to respect Ukraine's sovereignty and territorial integrity. They pledged not to use force or threats against the country.

Russia signed that document. The United States signed that document. The United Kingdom signed that document.

In 2014, Russia annexed Crimea and invaded eastern Ukraine. The security assurances turned out to be worth the paper they were printed on. In 2022, Russia launched a full-scale invasion that has killed hundreds of thousands of people and displaced millions. The country that gave up the world's third-largest nuclear arsenal because it trusted the guarantees of the major powers has been fighting for its survival ever since.

Ukraine's former officials have said publicly, repeatedly, and with undisguised bitterness that giving up the nuclear weapons was the worst decision in their country's history. Pavlo Rizanenko, a member of the Ukrainian parliament, told USA Today after the 2014 Crimea seizure: "We gave up nuclear weapons because of this agreement. Now, there's a strong sentiment in Ukraine that we made a big mistake." Zelenskyy himself, speaking at the Munich Security Conference five days before Russia's full-scale invasion in February 2022, tried three times to invoke the memorandum's consultation mechanism and was ignored each time. Even Bill Clinton has expressed regret. "I feel a personal stake because I got them to agree to give up their nuclear weapons," Clinton told Irish broadcaster RTÉ in 2023. "And none of them believe that Russia would have pulled this stunt if Ukraine still had their weapons." They're not wrong. If Ukraine had retained even a fraction of those warheads, Russia would never have invaded. Nuclear weapons are the one thing that makes a major power think twice about attacking you, and Ukraine gave them away for a promise.

. . .

North Korea looked at Libya and Ukraine and drew the obvious conclusion.

Kim Jong Il, and later his son Kim Jong Un, watched what happened to countries that abandoned their nuclear programs and countries that never had one. They watched the Americans praise Gaddafi and then bomb him. They watched the Americans guarantee Ukraine's security and then watch as Russia tore it apart. And they concluded, correctly, that the only reliable defense against American military power is a nuclear warhead sitting on top of a missile that can reach American territory.

North Korea endured decades of sanctions. Its people starved. Its economy was a hollow shell. But the Kim regime poured every available resource into the nuclear program, detonated its first device in 2006, tested progressively larger weapons, developed intercontinental ballistic missiles, and by the late 2010s had a deliverable nuclear capability that, however crude, changed the strategic calculus completely.

Nobody is bombing Pyongyang. Nobody is talking about regime change in North Korea. Instead, Donald Trump flew to Singapore in 2018 to shake Kim Jong Un's hand, stood next to him for photographs, and called him "very talented." The

80

leader of the world's most isolated dictatorship, running what amounts to a prison state, got a presidential summit because he had nuclear weapons. The leader of Libya, who gave up his nuclear program to please the West, got a bayonet.

The lesson is not subtle.

. . .

Now look at Iran through that lens.

In 2015, Iran signed the Joint Comprehensive Plan of Action, the JCPOA, commonly called the Iran nuclear deal. Iran agreed to limit its uranium enrichment, reduce its centrifuge stockpile, modify a heavy-water reactor, and accept intrusive international inspections. In exchange, the international community lifted nuclear-related sanctions and Iran regained access to the global financial system.

The deal was controversial. Critics, led by Netanyahu and Republican hawks in Congress, argued it was too lenient. Supporters argued it was the best achievable framework for preventing a nuclear-armed Iran without military action. Both sides had legitimate points. What's beyond dispute is that Iran signed it, and for several years Iran complied with its terms. The International Atomic Energy Agency verified compliance in report after report.

In 2018, Donald Trump pulled the United States out of the deal. Not because Iran had violated it. Because Trump didn't like it. He called it "the worst deal ever negotiated" and reimposed sanctions unilaterally. Iran, still technically in the deal with the remaining signatories, gradually resumed enrichment activities. The diplomatic framework that had constrained Iran's nuclear program the country that had championed nonproliferation destroyed it for fifty years.

And now, in 2026, the United States and Israel are bombing the country that signed the deal and kept it. The country that accepted limits on its program in good faith. The country that chose diplomacy over weapons and is now watching cruise missiles destroy its cities.

If you are an Iranian official, of any political stripe, in any future government that emerges from this war, what lesson do you draw?

There is only one rational answer. Get the bomb. Get it as fast as possible. Hide the program so deep underground that no bunker buster can reach it. Disperse it across so many sites that no air campaign can destroy all of them. And never, ever negotiate it away, because negotiation is a trap that ends with your cities in rubble and your leader in a drainage pipe.

And Iran is not the only country drawing this lesson. Saudi Arabia, which has been developing nuclear energy infrastructure with stated civilian intentions, has kept the enrichment option explicitly on the table. Crown Prince Mohammed bin Salman told CBS in 2018 that if Iran developed a nuclear weapon, Saudi Arabia would follow. Turkey has nuclear ambitions that it discusses openly. Egypt and the UAE have nuclear energy programs that could be redirected. Japan and South Korea, both covered by the American nuclear umbrella, have the technical capability to build weapons in months if they decide the umbrella is no longer reliable. The credibility of American security guarantees is the thread that holds together the entire nonproliferation regime. When the United States tears up a deal it signed, bombs the country that honored it, and proves through action that diplomacy does not protect you from American military power, that thread frays. And every country with the technical capability to build nuclear weapons quietly updates its calculations.

The nonproliferation regime that has prevented nuclear war since 1945 rests on three pillars: the NPT, which traded nuclear restraint for security guarantees; the IAEA, which verified compliance; and American credibility, which backed it all up. The Iran war has damaged all three. The NPT looks like a sucker's bet. The IAEA verified Iran's compliance and it didn't matter. And American credibility, already weakened by the withdrawal from the JCPOA, is now associated with bombing a country that did what America asked. The world that emerges from the Iran war will have more nuclear weapons in more countries than the world that went in. That is the strategic legacy of this conflict, and it will outlast every other consequence by decades.

• • •

Iran's strategy coming out of the ceasefire is so obvious it barely requires analysis.

Hold to the truce. Project the posture of a country willing to talk. Then negotiate. Accept whatever terms are on the table. Denuclearization framework, inspections regime, limitations, whatever the Americans and Israelis demand. Sign the paper. Smile for the cameras. Keep the bombing from resuming. Get the sanctions lifted. Start rebuilding.

And then do what North Korea did. What Pakistan did. What every country does when it decides the bomb is the only guarantee of survival.

Stall the inspectors. Build covert facilities in hardened mountain sites. Use the civilian nuclear infrastructure, which any deal will allow you to keep, as cover for parallel military research. Let the world's attention move on to the next crisis, which it always does. The Americans have the attention span of a fruit fly for foreign policy. Within two to three years, Iran will be yesterday's news. Within five, the inspectors will be filing complaints that nobody reads. Within ten, the centrifuges will be spinning in facilities the IAEA doesn't know about.

Pakistan ran this play for over a decade. A.Q. Khan built the bomb while Pakistan publicly denied everything, dodged inspections, and told the West what it wanted to hear. By the time the world acknowledged Pakistan had nuclear weapons, it was a fait accompli and nobody was going to do anything about it. India had done the same thing earlier. Israel did it before either of them, with the added twist of never officially admitting the weapons exist.

Iran has all of these models to study. It has the technical base, the scientific talent, and the industrial capacity. What it didn't have before was the political will, because there was a genuine internal debate about whether the costs of a weapons program outweighed the benefits. That debate is over. The strikes that hit Tehran resolved it in favor of the weapons faction, permanently.

• • •

The damage extends far beyond Iran.

Saudi Arabia has said for years, through multiple officials at multiple levels, that if Iran acquires nuclear weapons, Saudi Arabia will acquire its own. The mechanism is widely assumed to be Pakistan, which developed its nuclear capability with Saudi financial support and which maintains a relationship with Riyadh that goes beyond normal bilateral ties. Whether Saudi Arabia would build its own weapons or acquire them from Pakistan is an open question. That it would go nuclear is not. Saudi Crown Prince Mohammed bin Salman has said so explicitly.

Turkey has its own nuclear ambitions simmering under a surface of NATO membership and official compliance with the nonproliferation treaty. Erdogan has publicly questioned why Turkey shouldn't have nuclear weapons when other regional powers do. Turkey has the industrial and technical base to pursue a program if the political decision is made. An Iranian bomb and a Saudi bomb would almost certainly trigger a Turkish program.

South Korea and Japan have been having increasingly open domestic debates about independent nuclear deterrence. Both countries have the technical capability to build weapons quickly if they choose to. Both have relied on the American nuclear umbrella for decades. Both are watching the United States expend its military capacity in the Middle East while their region, the one that threatens them, gets less attention and fewer assets. If the American security guarantee starts to look unreliable, the logic of independent nuclear deterrence becomes compelling.

This is the proliferation cascade. One country's decision to go nuclear pressures its rivals and neighbors to follow. Each new nuclear state makes the next one more likely. The system that has kept the number of nuclear weapons states in the single digits for fifty years, the combination of the NPT, IAEA inspections, American security guarantees, and diplomatic pressure, is cracking under the weight of the demonstrations that the guarantees don't work and the diplomacy leads to bombing.

. . .

The implicit bargain of the nonproliferation regime was always this: you don't pursue nuclear weapons, and in exchange, the major powers don't use their conventional military superiority to bully you. You accept vulnerability as the price of membership in the international order, and the order protects you.

The United States just demonstrated, in the most graphic way possible, that the bargain doesn't hold. Iran accepted limits on its nuclear program through negotiation. The US withdrew from the agreement. Iran was bombed anyway. The message to every non-nuclear state is unambiguous:

compliance doesn't protect you. Negotiation doesn't protect you. Only nuclear weapons protect you.

That message will echo for decades. Long after the Iran war ends, long after the oil prices stabilize and the Strait of Hormuz reopens and the diplomats congratulate themselves on whatever settlement emerges, the proliferation consequences will be unfolding. More countries pursuing weapons. More fissile material in more places. More chances for accidents, miscalculations, theft, or use. A world with fifteen nuclear states is categorically more dangerous than a world with nine, and the Iran war may be the event that pushes us across that line.

This is the outcome that will matter most twenty years from now. Not the carrier that limped to Crete. Not the oil price spike. Not Netanyahu's political fortunes or Trump's approval ratings. The centrifuges. The ones spinning right now in programs nobody has announced yet, in countries that watched what happened to Libya, Ukraine, and Iran, and decided they would not be next.

. . .

Chapter 8: The War Nobody Can End

The United States entered the Iran war operating on a false assumption. The assumption was that the center of power in Iran was the civilian government: the presidency, the parliament, the diplomatic apparatus, the negotiating teams that had spent years working toward the JCPOA framework. Strike the leadership, destroy the nuclear infrastructure, and the civilian side would either collapse or moderate. That was the theory.

The theory was wrong on the first day.

The actual center of power in Iran was the Islamic Revolutionary Guard Corps. The IRGC answered not to the civilian government but directly to the Supreme Leader. It controlled Iran's missile program, its drone program, and its nuclear program, along with the proxy network that ran across Lebanon, Yemen, Iraq, and Syria, and it ran significant portions of the Iranian economy through its business conglomerates. When the US and Israel struck Iran on February 28, 2026, they were striking at a government that did not control the things they were trying to destroy. The IRGC assets, dispersed and hardened and deliberately separated from civilian structures, were the targets that mattered. The civilian government was the address the US had on file.

The result was predictable in retrospect. Striking the civilian government strengthened it domestically. Nothing unifies a population like an external attack on its institutions. And the IRGC's actual capabilities survived largely intact, because those assets aren't located at the addresses on file.

. . .

Ali Khamenei was killed on February 28, 2026, in his office compound, 86 years old, during the opening hours of the strike. The US had killed the Supreme Leader. This was supposed to be the decisive blow.

What followed demonstrated exactly why the IRGC framing matters.

Within hours of Khamenei's death, the IRGC pushed for an immediate succession, outside the legally prescribed procedures, before the Assembly of Experts could convene. They had a candidate ready: Mojtaba Khamenei, 56, the dead Supreme Leader's son, a man with strong IRGC ties, no official public office history, and no record of ever having to answer to anyone outside the circle. Trump had explicitly said Mojtaba would be an unacceptable choice. Netanyahu had said Israel would target anyone participating in the selection process.

Mojtaba was named Supreme Leader on March 8, 2026. One week after his father died.

Trump admitted that the US had ideas about preferred successors, but most of them died in the strikes. The attack, he told ABC News, had been so successful it knocked out most of the candidates. What he didn't say was the obvious implication: the IRGC had a candidate the US couldn't touch, and the succession happened anyway.

The regime that emerged from the opening strike was more hardline than the one the US attacked, controlled more completely by the IRGC, and led by a man whose legitimacy rests entirely on resistance to the United States. The US conducted a war against the wrong center of gravity and produced exactly the outcome it was trying to prevent.

. . .

Chapter 6 laid out the divergence between American and Israeli war aims. Trump wanted a manageable outcome he could deal with. Netanyahu wanted permanent degradation with no acceptable Iranian government at all. The succession settled that argument, and not in Netanyahu's favor or Trump's.

Trump wanted a workable successor government, sanctions relief tied to denuclearization, the Venezuela model applied to Iran. Netanyahu wanted a continued conflict that prevents any Iranian government from ever being in a position to threaten Israel. The IRGC and Mojtaba foreclosed both. There is no workable successor now. There is no Venezuela model available. There is only a more hardline government under IRGC control, and Netanyahu saying Israel is prepared to return to combat at any moment. The tail that wagged the dog got the war it wanted and an adversary it can't finish.

• • •

Here is what makes this war impossible to end cleanly. Count the actors and count their incentives.

The US military wanted sustainable operations and an exit. The magazine was running dry, the shipyards were behind schedule, and the force designed for short wars was fighting a long one.

US oil companies wanted the Strait closed and prices high. Chapter 10 worked through the mechanism: with energy production disrupted across the Persian Gulf, the United States became the world's largest oil exporter, Qatar's LNG production was suspended, US firms moved to fill the gap, and prices jumped more than 40 percent. That is not an incentive to end the war quickly.

US arms manufacturers wanted the war to continue. Every THAAD interceptor fired at twelve million dollars was a THAAD interceptor that needed replacing, and the same was true of every SM-6 and every Tomahawk. The companies that had lobbied against cheap drone swarms and directed-energy weapons because those would cannibalize their margins were the sole-source suppliers for the interceptors being consumed at rates they could not sustain, and that scarcity kept their prices and their contracts intact. A resolved Iran was their worst operating environment.

Israel wanted permanent degradation with no acceptable Iranian outcome short of that. Netanyahu said so explicitly after the ceasefire: there were still objectives to complete.

The IRGC and Mojtaba wanted exactly the prolonged conflict that validated their control. They had installed a Supreme Leader under wartime pressure, and his legitimacy depended on resistance. A negotiated peace in which Iran accepted limits would have ended everything the IRGC built in the succession crisis.

China wanted the shooting to stop. China's crude oil imports from the Gulf dropped 25 percent in March 2026, and economic pain drove Beijing's diplomacy. China provided the last-minute intervention that pushed Iran to accept the April 7 ceasefire. But while pushing for peace publicly, Chinese companies were supplying dual-use technology, missile parts and geospatial intelligence, to Iran. China emerged from the war as the indispensable peacemaker that Gulf states called when the US couldn't end its own conflict. Both Saudi Arabia and the UAE reached out to Xi Jinping directly in mid-April. The US built the crisis. China resolved it. That is a decade of American influence in the Gulf evaporating in four months.

Russia wanted the US bled but was too stretched in Ukraine to do more than provide satellite intelligence. US officials confirmed that Moscow supplied Iran with real-time data on American warships and aircraft, enabling more precise retaliatory strikes.

Saudi Arabia wanted the war to end, got hit anyway, secretly struck back, and concluded that the US security umbrella has holes in it. Iran struck Saudi oil refineries. Riyadh was targeted. Saudi Arabia intercepted the strikes and quietly conducted its own retaliatory airstrikes on Iranian drone and missile sites. The kingdom that normalized with Iran through a Chinese-brokered deal in 2023 ended the war's first phase buying interceptor missiles from Ukraine and reaching out to Beijing for security guarantees. The US-Saudi relationship has not ended. It has changed in a way that cannot be unchanged.

Every actor has incompatible objectives. The only thing they agree on is that the previous status quo is gone.

• • •

Chapter 7 made the proliferation case in full: Libya gave up its program and got a bayonet, Ukraine gave up its warheads and got invaded, North Korea kept the bomb and got a summit. The succession sharpened the lesson rather than changing it.

Khamenei was killed while Iran was complying with negotiated limits. For Mojtaba and the IRGC the calculation is not complicated. They are running a country that was attacked, had its Supreme Leader killed, installed a hardline successor under wartime pressure, and watched every negotiated agreement with the West proven worthless. Every day they don't have a bomb is a day they are vulnerable to the next strike Netanyahu is openly promising. Iran has no incentive now not to build one, and every country watching, Saudi Arabia and Turkey and Egypt and the UAE and South Korea and Japan, is running the same arithmetic Chapter 7 described. The war did not prevent proliferation. It made the case for nuclear weapons more compelling than any rogue state's defense ministry could have made on its own.

• • •

There is one more thread to pull, and it runs forward into the rest of this book.

The war did not cause the rebuilding. But it made the case for it impossible to ignore.

On April 3, 2026, the administration submitted a defense budget for FY2027 of $1.5 trillion, a 44 percent increase and the largest military budget proposal in US history. That number was not a war measure. The shipbuilding crisis it addresses predates the war by decades. It is the same hollowed-out industrial base documented in Chapters 1 and 2 and dissected in Chapter 20, the Newport News welders and the pipefitters and the nuclear-qualified technicians who were leaving the workforce faster than they could be replaced long before a single missile flew over the Persian Gulf. The budget would have been justified with or without Iran. The forces driving it are structural, not situational.

What the war changed was the politics. A defense establishment that cannot replace its interceptors, cannot build its ships on schedule, and cannot surge production when it needs to is an abstraction in a think-tank report and a live national emergency on a battlefield. The war turned the second into the first. The $65.8 billion for shipbuilding in that proposal is aimed squarely at the vocational pipeline that Chapter 21 describes. Not because the war created the need, but because the war made the need legible to people who could previously look away from it.

This is the mechanism behind the optimism this book promised, and it is worth being precise about what kind of optimism it is. Not the soft kind that says everything will work out. The hard kind. The tools exist. The money is being appropriated. The decision to act is being forced. Not by a politician who understood the structural forces, but by a war that made those forces impossible to ignore. That is a grim way to arrive at a fix. It is still a fix.

The trap has a shape. This chapter is part of its shape. And the exit, if there is one, runs through understanding exactly what the war exposed.

Part Two asked three questions. What are we doing? Who is driving? And what are the consequences?

The answers are bleak. We fought a war without defined objectives, driven by an ally whose goals diverged from ours, managed by a president whose decision-making ran on personal flattery rather than strategic analysis. And when it ended, the regime we attacked was more hardline, more completely controlled by the IRGC, and led by a man whose legitimacy rests on resistance to us. The consequence is the damaged nonproliferation regime that had kept the world from nuclear chaos for half a century, and a Middle East in which China, not the United States, is the power the Gulf states now call.

But these are symptoms, not causes. The reason the United States keeps stumbling into wars it can't explain is that it hasn't answered the fundamental question that the end of the Cold War left open: what is American power for?

For forty-five years, the answer was clear. Contain the Soviets. Everything flowed from that: the alliances, the bases, the fleet, the nuclear umbrella, the interventions, the entire global architecture. When the Soviet Union collapsed, the answer disappeared. And for thirty-five years, the United States has been using a military built for a purpose that no longer exists, deployed in conflicts that serve no coherent strategy, spending blood and treasure on wars that the public neither supports nor understands.

Part Three is about why. About the long retreat from global leadership that started before most Americans were born, the economic forces that drive it, and the question that matters more than any other: if the Strait of Hormuz isn't our problem, whose is it?

But here is the thing about wars without reasons: they can be ended. Ending them is a choice. Every carrier strike group freed from the Middle East is a carrier strike group available for the Pacific. Every dollar no longer burning in a desert is a dollar that can rebuild an industrial base. The situation described in Parts One and Two is grim. It is also reversible, if the country decides to stop doing the things that are making it weaker.

Part Three: The Long Retreat

Chapter 9: The Invoice Nobody Wants to Pay

The Iran war reached a ceasefire with a question still hanging over it. The United States fought to keep the Strait of Hormuz open, burned through a quarter of its interceptor stockpile doing it, and then watched the Gulf states call Beijing instead of Washington once the shooting paused. So whose strait was it? Whose oil, whose shipping lanes, whose problem? The United States imports almost none of the oil that moves through Hormuz. China imports most of it. America spent the blood and the missiles defending a chokepoint that matters far more to its chief rival than to itself. That is not an accident or a blunder. It is the residue of a bargain struck eighty years ago, one that most Americans have never heard of and are still paying for. To understand why the country keeps fighting wars like this one, you have to go back to the room where it agreed to.

In July 1944, while the war in Europe was still being fought, 730 delegates from 44 countries gathered at the Mount Washington Hotel in Bretton Woods, New Hampshire, to design the postwar economic order. The conference was dominated by two men: John Maynard Keynes, representing Britain, and Harry Dexter White, representing the United States. Over three weeks they hammered out the framework that would govern international trade and finance for the next eight decades: the International Monetary Fund, the World Bank, and a system of fixed exchange rates anchored to the US dollar, which was itself anchored to gold. White's design won. The dollar became the world's reserve currency, and America became the world's central banker.

It was the most consequential economic conference in human history. And it was, at its core, a transaction.

The United States, which would emerge from the war as the only major industrial economy still standing, agreed to underwrite the global system. Open markets. Free trade. Freedom of navigation on the seas. Security guarantees for allies. A nuclear umbrella for nations that agreed not to build their own weapons. The dollar as the world's reserve currency. American military bases on every continent. American ships on every ocean. American power, deployed globally, to maintain the order that let everyone trade and prosper.

In exchange, America got something worth more than any of those costs: a unified Western bloc that contained the Soviet Union. Every alliance, every base, every carrier patrol, every security guarantee served that purpose. NATO wasn't charity. It was the mechanism that kept Western Europe out of Soviet hands. The Japan alliance wasn't altruism. It was the anchor that kept the Pacific from becoming a Soviet lake. The entire system existed because containing Communism required a global coalition, and a global coalition required someone to organize, fund, and defend it.

The system worked. The Soviet Union was contained, then exhausted, then dissolved. The Cold War ended without nuclear exchange. The liberal economic order produced the greatest period of sustained global prosperity in human history. By any measure, the American investment paid off spectacularly.

Then the wall came down. And the invoice kept coming.

• • •

November 9, 1989. The Berlin Wall falls. Within two years, the Soviet Union ceases to exist. The strategic rationale for every alliance, every base, every carrier patrol, every security guarantee evaporates overnight.

Americans didn't maintain a 600-ship Navy and garrison Europe out of generosity. They did it because the alternative was Soviet dominance of Eurasia. When that threat disappeared, the entire cost structure of global leadership became optional. Not immediately, not dramatically, but fundamentally. Every dollar spent maintaining the global order was now a dollar not spent on domestic priorities, and without an existential threat to justify the expense, the political case for paying it collapsed.

The retreat started almost immediately, though nobody called it that at the time. Bill Clinton won the 1992 election partly on a domestic platform: "It's the economy, stupid." The peace dividend wasn't a strategic concept. It was a domestic spending concept. The Cold War is over. We won. Now take the defense budget and spend it on things voters care about.

The Navy went from 594 ships to under 350 during the Clinton years. The Army shrank by a third. Bases closed across the country. The defense industrial base consolidated from dozens of major contractors to five or six, as companies merged or went under because there weren't enough contracts to sustain them all. The intelligence community was cut. The State Department was starved of resources. The entire infrastructure of global engagement was downsized on the theory that history had ended and the future would take care of itself.

The theory had a name: the unipolar moment. One superpower, no serious challengers, a world converging toward liberal democracy and free markets. Francis Fukuyama published "The End of History" and the political establishment treated it as prophecy rather than thesis. Free trade would democratize China. Russia would liberalize. The Middle East would sort itself out. American military dominance was so overwhelming that no rational actor would challenge it, so the military could safely shrink because it would never need to fight a real war again.

Every one of those assumptions turned out to be wrong. But the drawdown continued anyway because the domestic political incentive had flipped permanently.

• • •

The conventional narrative says the retreat from global leadership started with Trump. That's the comfortable version because it lets everyone else off the hook. Reality is that every president since 1991 has been withdrawing from the global order. They just did it at different speeds, with different rhetoric, and with varying degrees of self-awareness.

Clinton cut the military and focused on domestic economics. He intervened selectively, in Bosnia and Kosovo, but only when European allies begged and only with air power that minimized American casualties. He treated the Middle East as a diplomatic problem rather than a military one, which was the correct instinct at the time. But he also let the defense industrial base hollow out, let the intelligence community atrophy, and let the diplomatic corps wither. The infrastructure of global engagement was dismantled during the decade when it was cheapest to maintain.

The scale of the commitment being abandoned deserves a moment of clarity, because most Americans have no idea what they were paying for. The United States operates roughly 750 military bases in at least 80 countries. It maintains carrier strike groups in the Pacific, the Mediterranean, and the Persian Gulf on a continuous rotation. It provides the nuclear umbrella that protects 30 NATO allies plus Japan, South Korea, and Australia. It guarantees freedom of navigation on every ocean. It funds international institutions from the World Bank to the UN peacekeeping budget. It provides bilateral military aid to over a hundred countries.

The total cost of this system, direct and indirect, runs to roughly $1.5 trillion per year when you include the defense budget, intelligence community, Veterans Affairs, nuclear weapons maintenance, and the interest on debt accumulated from prior military spending. That is more than the federal government spends on education, transportation, science, and the environment combined.

Americans were paying for a global empire and receiving, in return, a world that mostly worked. When the empire started to retract, the world started to stop working. But the bill didn't get any smaller, because the commitments lingered even as the strategic rationale evaporated.

George W. Bush looked like a reversal. Massive military spending. Two wars. The Global War on Terror as an organizing principle that seemed to replace containment as the rationale for global presence. But Iraq and Afghanistan accelerated the retreat in a different way. They consumed a trillion dollars in direct costs and several trillion more in long-term obligations. They burned through the military's equipment, personnel, and morale. They destroyed public appetite for intervention. And they demonstrated, to the world and to the American public, that the United States could not achieve its stated objectives through military force. The overreach didn't rebuild the order. It discredited the idea of maintaining it.

Obama made the pivot to Asia explicit, which was an admission that the US couldn't be everywhere. The "rebalance to the Pacific" was strategic shorthand for "we're going to stop spending so much on the Middle East and start paying attention to China." It was the right instinct. It never fully materialized because the Middle East kept pulling resources back. Syria. ISIS. The Iran nuclear deal, which was an attempt to solve the Iran problem diplomatically so that military resources could be redeployed eastward. When Trump pulled out of the deal, he didn't just kill the agreement. He killed the strategic logic behind it.

Trump 1.0 said the quiet part loud. NATO allies are freeloaders. Trade deals are bad. Why are we defending South Korea? Why are we in Syria? The rhetoric was shocking to the foreign policy establishment, but the instinct wasn't new. It was the logical endpoint of a twenty-five-year trend. Every predecessor had been asking the same questions behind closed doors. Trump asked them on Twitter.

Biden tried to reverse the rhetoric without reversing the substance. He talked about alliances and the rules-based order while continuing the Afghanistan withdrawal, slow-walking Ukraine support through bureaucratic caution, and failing to rebuild the industrial base or the fleet. The gap between Biden's speeches and Biden's actions was the gap between what the establishment wanted to believe and what the country was willing to do.

Trump 2.0 is running a hot war while also undermining the alliance structure that's supposed to support it. Gulf allies initially blocked US basing access because they didn't trust the commitment. NATO is involved tentatively. The "order" that's supposed to enable American power projection is fraying at every seam while American power is being projected at maximum intensity.

• • •

The structural explanation for all of this is simpler than the political narrative suggests.

The United States is a continental power. Two oceans. Two weak neighbors. More natural resources than any country on the planet. The Mississippi River system, which is the single greatest geographic advantage any nation has ever possessed, provides cheap bulk transportation from the interior to the coast across the most productive agricultural land on Earth. Everything a civilization needs to be secure, prosperous, and self-sufficient is contained within the borders or immediately next door.

The natural posture of such a country is hemispheric dominance. Control your continent. Keep potential rivals from establishing a foothold in your hemisphere. And let the rest of the world sort itself out.

That was American strategy from 1823, when James Monroe declared the Western Hemisphere off-limits to European colonial expansion, until 1941, when Pearl Harbor forced the country into global engagement. For 118 years, the United States did exactly what its geography suggested: it dominated its hemisphere and stayed out of other people's problems.

The period of global hegemony, from 1941 to the present, is the anomaly. Not the norm. It was forced by two specific historical circumstances: World War II, which required global engagement to defeat fascism, and the Cold War, which required global engagement to contain Communism. Both of those circumstances are gone. What we're watching, across all six presidencies since 1991, is the reversion to the historical mean. Not a choice. A gravitational pull.

George Friedman, the geopolitical analyst who has influenced more of this book's thinking than any other single source, puts it bluntly: the United States was never meant to be a global hegemon. It was pulled into that role by circumstances and it's being pulled back out by the absence of those circumstances. Every president is managing the reversion, whether they know it or not. The ones who resist it, like the neoconservatives under Bush, make things worse through overreach. The ones who accelerate it, like Trump, make things worse through chaos. The optimal path is managed withdrawal: intentional, strategic, and calibrated to maintain essential interests while shedding commitments that no longer serve them.

Nobody has managed it well. The withdrawal has been stumbled through, not planned. Wars of choice in the Middle East consumed the resources that should have funded the transition. The alliance structure was allowed to deteriorate through neglect rather than being deliberately restructured. The industrial base was hollowed out. The fleet shrank. The diplomatic corps withered. And now the country is fighting another Middle Eastern war that it didn't plan for, using military assets it can't replace, while the adversary it should be focused on builds railroads across Asia.

• • •

The question that nobody in Washington wants to answer honestly is this: if the US is going to withdraw from global hegemony, what does it keep and what does it let go?

The Western Hemisphere is non-negotiable. That's the Monroe Doctrine and it's as valid today as it was in 1823. North and South America, with their combined resources, population, agricultural capacity, and geographic protection, are the natural sphere of American dominance. Keeping potential rivals, particularly China, from establishing military or economic footholds in the hemisphere is a genuine vital interest.

The Pacific is the real debate. China's rise makes the Western Pacific the most strategically consequential region on the planet. Whether the US maintains its Pacific alliance structure, keeps its bases in Japan and South Korea and Guam, and continues to guarantee Taiwan's security is the defining strategic question of the next fifty years. There are strong arguments on both sides, and this book will explore them in Parts Five and Seven.

The Middle East is the easy one, and it's the one nobody will say out loud. America does not need the Middle East. It doesn't need the oil. It doesn't need the bases. It doesn't need the alliances. Every dollar spent, every life lost, every ship deployed in the Persian Gulf is a resource subtracted from somewhere that matters.

The Iran war is the most expensive proof yet of a proposition that's been true for at least a decade: the Middle East is somebody else's problem.

The proof is in the strait. Eighty-four percent of the oil that flows through the Strait of Hormuz goes to countries that aren't the United States. American sailors are dying to protect someone else's supply chain.

. . .

So if the invoice is the problem, here is where I would start tearing it up. We are throwing too much at advanced research programs that will not be ready when we need them and not enough at building the inventory we already know how to build. Counter China where we need to. Spend the rest on ships, aircraft, and ammunition that exist and work. What we cannot crew or deploy right now should go into mothballs in the desert, maintained and ready to come back fast. Build ships specifically designed to be stored and reactivated. The lesson of World War Two is that the country that can produce fastest wins. We have traded surge capacity for prestige projects, and that is exactly backwards.

Chapter 10: Somebody Else's Strait

The Strait of Hormuz is 21 miles wide at its narrowest point. It separates Iran from Oman. Through it passes roughly 20 million barrels of oil per day, about 20 percent of the world's seaborne oil trade. On a normal day, dozens of supertankers glide through its shipping lanes carrying the lifeblood of the global economy from Persian Gulf producers to customers in Asia, Europe, and everywhere else.

When the Iran war started on February 28, 2026, Iran effectively closed it. The IRGC announced that the strait was shut to vessels from the US, Israel, and their allies. At least five tankers were damaged. Two crew members were killed. About 150 ships were stranded. Maritime traffic dropped to a trickle. Brent crude surged past $100 a barrel for the first time in four years, peaking at $126. The International Energy Agency launched the largest emergency reserve release in its history, 400 million barrels, to prevent the global economy from seizing up.

Oil prices in America spiked. Gas approached four dollars a gallon. President Trump went on Truth Social and said that the closure "doesn't really affect" the United States because "we have so much oil."

He was half right. And the half he was wrong about is the most important economic and strategic fact in this entire book.

. . .

The United States produces roughly 13 million barrels of oil per day. That's more than any other country on the planet. More than Saudi Arabia. More than Russia. American shale revolutionized the global energy market and turned the US from a net importer into the world's largest producer. On paper, the United States is energy independent. It produces more than it consumes.

So why did American gas prices spike when a strait 7,000 miles away closed?

Because oil is a globally fungible commodity priced on world markets. American oil companies don't sell their product at a special American price. They sell it at the world price, because the world price is higher and they're in business to make money, not to subsidize American consumers out of patriotism. When the Strait of Hormuz closes and 20 percent of global supply is disrupted, the world price goes up. When the world price goes up, American prices go up too, even for oil pumped out of the ground in West Texas.

The American consumer pays more. The American oil company makes more. The spread between the cost of extracting Permian Basin crude, roughly $40 to $50 a barrel, and the world price of $126 a barrel is pure windfall profit. And that windfall is generated by a war that American taxpayers are also paying for through their defense budget.

Read that again. American oil companies profit from the war that American taxpayers fund. The taxpayer gets the bill twice: once for the missiles and the ships and the sailors, and again at the gas pump. The oil company gets paid twice: once from the elevated world price, and again from the contracts to fuel the military operations that created the price spike.

This is not a conspiracy. It's a market structure. And the market structure is a policy choice.

The natural gas market makes the perverse incentives even worse. The United States is now the world's largest exporter of liquefied natural gas. When the Strait of Hormuz closed and Qatari LNG stopped flowing, Asian and European buyers scrambled for alternatives. American LNG became the most valuable commodity on the energy market. Export terminals in Louisiana and Texas were running at maximum capacity, selling gas at prices that would have been unimaginable a month earlier. The energy industry doesn't just profit from oil price spikes during Hormuz crises. It profits from gas price spikes too. Every disruption to Middle Eastern energy flows is a revenue event for American exporters, which means the industry has a structural financial interest in the instability that American soldiers are deployed to manage. Nobody in Houston is rooting for war. But nobody in Houston is losing money on it either.

. . .

Now look at who depends on the Strait of Hormuz.

Eighty-four percent of crude oil and condensate shipped through the strait goes to Asian markets. China gets about a third of its oil through Hormuz. Japan gets roughly 95 percent of its crude from the Middle East, most of it through the strait. South Korea gets over 70 percent. India is a massive and growing customer.

Europe gets much of its jet fuel and about 12 to 14 percent of its liquefied natural gas from Qatar, shipped through the strait.

The United States gets about 490,000 barrels per day from Persian Gulf countries. That's a fraction of its total consumption and a tiny fraction of what flows through Hormuz. In a pinch, the US could replace those barrels from domestic production, Canadian imports, or other sources. It would be inconvenient. It would not be catastrophic.

So the question becomes: why are American sailors standing watch over a shipping lane that primarily serves China, Japan, South Korea, India, and Europe? Why are American missiles being expended to keep it open? Why are American taxpayers footing the bill for defending a supply chain that benefits other countries far more than it benefits the United States?

The answer is inertia. And the inertia has a history.

• • •

The American commitment to Persian Gulf security dates to 1979. The Iranian Revolution, followed by the Soviet invasion of Afghanistan, raised the specter of hostile powers controlling the oil supply that the Western world depended on. President Jimmy Carter declared in his 1980 State of the Union address that any attempt by an outside force to gain control of the Persian Gulf region would be regarded as an assault on vital American interests and repelled by any means necessary, including military force.

That was the Carter Doctrine, and it made sense in 1980. The United States was a net oil importer, heavily dependent on Middle Eastern crude. A disruption of Persian Gulf oil would have crippled the American economy. The Soviet Union was a potential aggressor. Defending the Gulf was defending American prosperity.

The problem is that the world changed and the policy didn't. American oil production doubled over the past fifteen years. The shale revolution made the US energy-independent in net terms. The Soviet Union collapsed. The original rationale for the Carter Doctrine, defending American oil imports from Soviet aggression, ceased to exist on both sides of the equation. The US doesn't need the oil, and the Soviets aren't threatening it.

But the infrastructure of intervention remained. The bases in Bahrain, Qatar, Kuwait, and the UAE. The Fifth Fleet. The carrier patrols. The air wings. The prepositioned equipment. Tens of thousands of American military personnel permanently stationed in a region that the US no longer depends on for its economic survival. The inertia of a commitment made when the circumstances were different, maintained because nobody ever formally decided to stop.

Each crisis reinforced the commitment rather than questioning it. The Gulf War in 1991 seemed to validate the Carter Doctrine. The Iraq invasion in 2003 extended the commitment. The Obama-era pivot to Asia was supposed to start the rebalancing, but the Middle East kept dragging resources back. And now the Iran war has pulled the largest American military deployment to the region since 2003, at a time when the strategic argument for being there has never been weaker.

* * *

The strategic absurdity reaches its peak when you consider the country that benefits most from American protection of Hormuz.

China. The country the United States is supposed to be in strategic competition with. The country that the entire Pacific force posture is designed to deter. The country that every National Defense Strategy identifies as the primary threat.

China gets far more of its energy through the Strait of Hormuz than the US does. And right now, American ships and missiles are being depleted to keep China's oil flowing, while China sits on the sidelines, sells Iran anti-ship missile technology, provides BeiDou satellite navigation for Iranian weapons, and takes notes on American munitions consumption rates and carrier vulnerabilities.

The United States is subsidizing its primary competitor's energy security while degrading its own military readiness for the competition that matters. That's not strategy. That's insanity.

Japan, South Korea, India, Europe: these are wealthy, technologically advanced nations with the capacity to build navies, patrol shipping lanes, and defend their own energy supply chains. They choose not to because the United States has done it for them for forty-five years. The Hormuz patrol is the purest example of the free-rider problem in the entire global order. America provides the security. Everyone else reaps the benefit. The American taxpayer gets the bill plus the blowback from every conflict that erupts in the region.

• • •

The solution is obvious, and it has two parts.

First, decouple American oil prices from the global market. This is the hard part, because it requires fighting the oil industry. Several mechanisms could work. Export controls: the US banned crude oil exports from 1975 to 2015. The ban could be reimposed, or a partial version could require domestic demand to be met at a capped price before exports are allowed. Strategic pricing mechanisms: a domestic price floor to keep extraction profitable and a ceiling to prevent global shocks from hitting the pump, with the government absorbing the spread. Mandatory reserve requirements: force domestic producers to maintain a certain volume in domestic storage before exporting.

Every one of these approaches faces opposition from the oil industry, free-market ideologues, and trading partners who depend on US exports. The oil companies will spend billions fighting it because the current system is perfect for them: global crises spike prices, they make windfall profits on domestic production that costs the same to extract regardless of what's happening in the Persian Gulf.

But the strategic logic is overwhelming. If the US had a decoupled domestic energy market right now, the Strait of Hormuz closure would be somebody else's problem. Gas would be $2.50. The economic impact would be minimal. And the political pressure to get involved in Middle Eastern wars would evaporate, because why would American voters support spending billions and losing soldiers to keep open a shipping lane that doesn't affect their gas prices?

Second, tell the countries that depend on Hormuz to defend it themselves. Japan, South Korea, India, Europe: build the navies, deploy the escorts, negotiate the deals. The US will sell weapons if they want them. But the days of bleeding for other people's oil are over.

This would accomplish several things at once. It would force allied nations to invest in their own defense, something the US has been begging them to do for decades. It would free the Navy for the Pacific, where the strategic competition matters. It would remove the primary justification for American military involvement in the Middle East. And it would put China in an interesting position: suddenly they're the ones who need Hormuz open, and they're the ones who have to figure out how to secure it.

• • •

The counterargument is always "but the global economy" and "alliance cohesion." And yes, pulling out of Hormuz overnight with no transition would be chaotic. But the current system, where the US spends blood and treasure protecting a supply chain that primarily benefits other countries while those countries lecture Washington about its foreign policy, is not sustainable either. This war is proving that.

The shale revolution should have been the trigger for a fundamental strategic reorientation. The US achieved energy independence and then did absolutely nothing to translate that into strategic independence. Instead it kept the same force posture, the same alliance commitments, the same entanglements. And now it's fighting a war in 2026 over a shipping lane it doesn't need, while the countries that do need it watch from the sidelines.

Energy independence is meaningless without price independence. The US produces enough oil but chose to stay integrated into global pricing because it was profitable for the industry. The consequence is a self-reinforcing loop: global crisis spikes prices, which justifies intervention, which creates more crises, which spikes prices further. The oil companies profit on both ends. The taxpayer pays on both ends. The soldiers die in the middle.

Breaking that loop would be the single most strategically important thing the United States could do for its long-term security. More important than any weapons system. More important than any alliance. More important than any war. If Hormuz doesn't affect American gas prices, the entire strategic rationale for maintaining a massive naval presence in the Persian Gulf collapses. And that frees the ships, the munitions, and the attention for the Pacific, which is where the actual strategic competition of this century is playing out.

But it requires a president willing to fight the oil industry. And we haven't had one of those yet.

• • •

The Strait of Hormuz is somebody else's problem. It has been somebody else's problem for at least a decade, probably longer. America has been paying for it out of inertia, habit, and the financial interests of an oil industry that profits from the arrangement. The cost is measured in ships worn out, missiles expended, soldiers killed, and attention diverted from the threat that matters.

But the Strait is just the most visible symptom of a deeper condition. The entire American posture in the Middle East, and increasingly in Europe and other regions, is a legacy commitment maintained past its expiration date. The global order the US built after World War II was worth the cost when it served a clear purpose. That purpose is gone. The invoice keeps coming. And the country is running out of capacity to pay it while also preparing for the challenges that will define the next fifty years.

Those challenges are bigger than any war. Bigger than any strait. Bigger than any single adversary. And they're already here.

But first, the war that's already underway. The one that most people don't recognize because they're looking for the wrong shape.

Chapter 11: The War That's Already Here

Most Americans are not tracking any of this. They are focused on one question: can I afford my lifestyle? That is not a criticism. It is how most people have always operated. The geopolitical picture only breaks through when the economic pain gets bad enough. When it does, they vote in a new party. The problem is that by the time the pain from a non-kinetic war becomes visible at the grocery store and the gas pump, the adversary has already been operating for years. The war that is already here does not announce itself. It shows up as higher prices, unreliable infrastructure, and institutions that stopped working for reasons nobody can quite explain.

In 2026, a pollster asked Americans how likely they thought it was that a world war would break out in the next four years. Sixty-three percent said somewhat or very likely. Among voters under 50, it was 75 percent.

They were answering the wrong question. It wasn't going to break out. It already had.

• • •

When people hear "world war," they picture 1941. Formal declarations. Massed armies. Tank columns rolling across borders. Carrier fleets steaming toward each other. Sides clearly drawn: Allies versus Axis. A starting gun, a climactic battle, a surrender ceremony on the deck of a battleship. That's the shape they're looking for, and since they don't see it, they conclude it hasn't happened yet.

But wars don't always announce themselves. World War I didn't start with a grand strategic decision. It started with an assassination in Sarajevo that triggered a cascade of alliance obligations, miscalculations, and escalations that dragged a continent into the slaughter over the course of weeks. Nobody planned the Western Front. It emerged from a series of events that, individually, seemed manageable and, collectively, produced catastrophe.

Look at the current map the same way. Not as a series of separate crises, but as a single interconnected system of conflicts involving major powers, spanning multiple continents, with global economic consequences. Then ask yourself whether the shape you're looking at is recognizable.

. . .

February 2022. Russia invades Ukraine. A full-scale land war in Europe for the first time since 1945. A nuclear-armed great power attacking a sovereign European nation, in direct violation of the international order the US built after World War II. NATO countries respond with massive arms shipments, intelligence sharing, economic sanctions, and training programs that make them co-belligerents in everything but name. Russia's professional military is shattered over two years of attritional combat. Hundreds of thousands dead on both sides. The war is still going in 2026 with no resolution in sight.

That alone would be the most significant military conflict in Europe in eighty years. But it's not alone.

October 2023. Hamas attacks Israel. The subsequent Israeli campaign in Gaza kills tens of thousands. The conflict expands. Hezbollah opens a front from Lebanon. The Houthis begin attacking commercial shipping in the Red Sea, disrupting one of the world's most important trade routes. The eastern Mediterranean becomes a combat zone. American warships deploy to defend against drone and missile attacks.

June 2025. Israel and the US strike Iranian nuclear facilities in a twelve-day air campaign. The first direct American military action against Iran. A preview of what's coming.

February 2026. The US and Israel launch full-scale strikes against Iran. Iran retaliates against US bases across the Middle East, against Gulf state allies, against Israel. The Strait of Hormuz closes. Oil prices spike. The conflict spreads to Lebanon, Iraq, Bahrain, the UAE, Qatar, and Saudi Arabia. Missiles hit civilian targets across a dozen countries. Over 2,300 people killed in the first month.

Add them up. A land war in Europe. A multi-front conflict in the Middle East. A maritime war in the Red Sea. A naval conflict in the Persian Gulf and Arabian Sea. Major powers involved directly or through proxies. Global energy markets disrupted. Supply chains broken. Alliance structures strained. Nuclear threats brandished. The geographic scope spans from Ukraine to Yemen, from the Mediterranean to the Indian Ocean.

If that's not a world war, the definition needs updating.

. . .

The reason people don't recognize it is that it doesn't have the features they associate with the term. There's no formal declaration. The sides aren't cleanly drawn. The major powers aren't fighting each other directly. Russia is fighting Ukraine while NATO arms Ukraine. The US and Israel are fighting Iran while China provides Iran with technology and intelligence. Nobody has declared war on anybody else. The conflicts are nominally separate: the Ukraine war, the Gaza war, the Iran war. Different theaters, different combatants, different causes.

But they're connected. Russia and Iran cooperate on weapons: Iranian drones in Ukraine, Russian air defense systems in Iran. China provides technology and intelligence to Iran while buying Russian energy at a discount. The Houthis, an Iranian proxy, attack commercial shipping to pressure the Western coalition. Hezbollah, another Iranian proxy, opens a second front against Israel. Each conflict feeds the others. Each escalation in one theater changes the calculus in another.

Russia, China, and Iran aren't formally allied in the World War II Axis sense. They don't have a mutual defense treaty or a unified command structure. What they have is something that might be worse: a flexible, deniable partnership of convenience. They share intelligence when it suits them. They sell each other weapons when it's profitable. They coordinate diplomatically when it serves their interests. And they collectively stress-test the American security umbrella at multiple points at the same time, forcing the US to divide its attention and resources across theaters that are all demanding at once.

The US is providing weapons to Ukraine, defending Israel, bombing Iran, protecting Gulf states, patrolling the Red Sea, and maintaining deterrence in the Pacific. With a military that's been shrinking for thirty years and an industrial base that can't produce munitions fast enough for one of those commitments, let alone all of them.

The economic dimensions are as significant as the military ones and harder to see. Russia weaponized energy exports against Europe in 2022, cutting gas supplies to punish countries that supported Ukraine. The West responded with financial sanctions that froze $300 billion in Russian central bank reserves and cut Russian banks off from the SWIFT payment system. China retaliated against Lithuania for opening a Taiwanese representative office by blocking Lithuanian exports through informal customs barriers. The

United States restricted semiconductor exports to China, aiming to slow Chinese AI development by a generation. China restricted exports of gallium and germanium, critical minerals used in chip manufacturing. Each move produced countermoves. Each economic weapon revealed new vulnerabilities. The global economy, which was supposed to make war irrational through interdependence, is being carved into competing blocs where trade relationships double as pressure points and supply chains double as weapons. The economic warfare is quieter than the missile strikes, but it may be more consequential. Wars end. Trade regimes, once fractured, take decades to rebuild.

The cyber dimension runs underneath everything else. Russian state hackers have attacked Ukrainian power grids, American pipelines, European hospitals, and election systems across the democratic world. Chinese cyber operations have penetrated American defense contractors, stolen terabytes of classified data on weapons systems including the F-35, and pre-positioned malware in American critical infrastructure that could be activated during a future conflict. Iran has conducted cyberattacks against Saudi oil facilities and American financial institutions. None of these operations rise to the level of an act of war under traditional definitions, which is the point. They exist in the gray zone between peace and war, inflicting damage and gathering intelligence without triggering the kind of response that would escalate to open conflict. But they are warfare. They serve military objectives. They weaken adversaries. They prepare the battlefield. And they are happening continuously, across all the theaters described above, on a scale that makes the Cold War's espionage efforts look quaint.

The information dimension may be the most consequential of all, because it targets the decision-making capacity of democratic societies directly. Foreign information manipulation and interference, known as FIMI in intelligence circles, does not try to convince populations of a specific lie. It floods the information environment with so many contradictory narratives, some true, some false, most unverifiable, that the cost of determining what is real exceeds what most people are willing to pay. The cost-exchange ratio from Chapter 2 applies here too: fabricating a narrative costs almost nothing, but disproving it requires research, sourcing, and time that the news cycle does not allow. By the time a debunking is published, the original claim has already been shared a million times and the audience has moved on. The informational equivalent of a hundred-to-one missile ratio, and just as devastating to the defender. The goal is not belief in a falsehood. The goal is the erosion of belief in any truth.

Russia has been running these operations for over a decade, and the results are visible across the democratic world. In late 2024, a Romanian presidential candidate named Calin Georgescu surged from obscurity to the final round of the presidential election on the back of a massive foreign-sponsored digital operation, forcing the suspension of the democratic process entirely. In Hungary, Kremlin-linked operatives developed campaigns to support Viktor Orban's reelection through Russian-designed posts and videos laundered through local influencers to appear organic. Orban, the most pro-Russia leader in the EU, has repeatedly blocked sanctions on Russia and reportedly passed information from European Council meetings to Moscow. The campaigns work because democracies are structurally vulnerable to them: open information systems are easy to penetrate, free speech norms prevent censorship of foreign-origin content, and regular elections provide predictable high-value targets. Authoritarian regimes face none of these constraints. They control their domestic information environments, suppress dissent, and launch operations against democracies from behind a one-way mirror.

The Iran war has its own information battlefield. Iranian state networks flood platforms with fabricated battlefield outcomes and exaggerated casualty figures. Israeli and

American messaging focuses on justifying military action. Gulf states suppress images of Iranian missile damage to maintain the appearance of insulation from the conflict. Every belligerent is fighting for the narrative, because in democracies, the narrative shapes the policy, and the policy shapes the war.

• • •

The nuclear dimension is the one that keeps the 63 percent up at night, and their instinct is both right and wrong.

Right in that nuclear weapons are part of this conflict. Russia has made explicit nuclear threats over Ukraine. Putin revised Russia's nuclear doctrine to lower the threshold for use. The war in Iran involves strikes against nuclear facilities by both sides. Iran's entire nuclear program is both a cause and a consequence of the conflict. North Korea, a nuclear state, is providing artillery shells to Russia. The nuclear dimension is woven through every theater.

Wrong in that the most likely outcome is not a strategic nuclear exchange. The mushroom-cloud-over-New-York scenario is not what this war is heading toward, because it would destroy everyone involved and everyone knows it. Mutual assured destruction still works as a ceiling on escalation. It worked during the Cold War against an adversary far more dangerous than any current one. It's working now: Russia invaded a European country, got hammered, lost hundreds of thousands of troops, and didn't go nuclear. The taboo held.

Theater nuclear weapons are a different story. A tactical nuke against a hardened underground facility, a massed military formation, or a carrier strike group in deep ocean is thinkable in ways that a strategic exchange is not. The weapon is smaller. The target is military. The fallout is contained, or at least containable. The escalation logic is different: you're trying to shock the adversary into stopping, not destroy their civilization.

Russia's doctrine explicitly includes the concept of "escalate to de-escalate": use a limited nuclear strike to freeze a conflict on favorable terms, on the theory that the other side will back down rather than risk full exchange. China has the DF-26, a missile that can carry either a conventional or nuclear warhead and that was designed to threaten carrier strike groups. A nuclear-tipped DF-26 hitting a carrier in deep ocean, far from population centers, would destroy a major military asset while keeping the fallout literally in the middle of the sea. The US would have to decide whether to escalate to strategic exchange over a military target at sea. That's a decision nobody wants to make, which is exactly the point.

The danger of theater nukes isn't the blast. It's that nobody has a proven playbook for what happens after one gets used. Every war game and simulation shows the same problem: once one side crosses the nuclear threshold, the escalation ladder gets very slippery. The side that got hit faces enormous pressure to respond in kind. The side that used it has already broken the taboo, making the second use psychologically easier. The theoretical firebreak between tactical and strategic nuclear use exists on paper but has never been tested in reality.

The Cold War's nuclear stability rested on two players with roughly symmetrical arsenals and decades to develop mutual understanding of red lines. The emerging environment has no such clarity. Russia, China, and the United States are three nuclear peers for the first time in history, with China expanding its arsenal from roughly 350 warheads toward an estimated 1,000 by 2030. Add the regional nuclear states: Israel, Pakistan, India, North Korea, and potentially Iran and Saudi Arabia within a decade if the proliferation cascade from Chapter 7 materializes. Each additional nuclear state creates new escalation pathways that don't run through Washington or Moscow. A Pakistan-India exchange triggered by climate-driven water conflict. An Israeli strike on a nascent Iranian device. A North Korean miscalculation during a Korean Peninsula crisis. Each is low probability in any given year. Across thirty years, with nine or ten nuclear states and active conflicts involving several of them, the cumulative probability of at least one nuclear use is not a rounding error. It is the background radiation of the era this book describes.

The most likely path for this war, and for the broader conflict it's part of, is that it stays conventional. Not because anyone chooses restraint out of principle, but because the nuclear arsenals create a ceiling that nobody wants to test. The conflict stays below that ceiling, which means it grinds. Long, attritional, multi-theater, economically devastating. Fought over energy and shipping lanes and industrial capacity rather than territory. More like World War I than World War II in character, except spread across the globe and with cyber and economic warfare doing a lot of the damage that artillery did in the trenches.

· · ·

The cyber and economic dimensions are the parts of this world war that don't show up on maps but may matter more than the kinetic fighting.

Sanctions against Russia constituted the most aggressive economic warfare campaign since World War II. Russia retaliated by weaponizing energy supplies to Europe. Iran's closure of Hormuz is economic warfare at the global scale, removing 20 percent of seaborne oil supply to coerce the coalition into stopping. Oil price spikes function as a tax on every economy that imports energy, transferring wealth from consumers to producers and destabilizing governments that depend on affordable fuel to keep their populations from rioting.

Cyber attacks are the constant background noise. Russian attacks on Ukrainian infrastructure. Attempts to disrupt Western financial systems. Chinese penetration of defense networks. Iranian attacks on Gulf state infrastructure. None of this produces the dramatic footage of a missile strike, but the cumulative damage to trust, infrastructure, and economic stability is enormous and ongoing.

The information war runs parallel to everything else. Russian propaganda operations across Western democracies. Chinese influence campaigns. Iranian state media. Each combatant is fighting for the narrative as aggressively as they're fighting for territory or sea lanes. The American domestic debate about the Iran war, with its confusing poll numbers and unclear objectives, is itself partly a product of information warfare designed to erode public support for the conflict.

• • •

The deeper pattern underneath all of this is the one that the structural geopoliticists, Friedman chief among them, have been describing for years. The post-1945 American-led order is fracturing along multiple fault lines at once. The fracturing isn't caused by any single event or any single adversary. It's caused by the withdrawal of the force that held the system together: American commitment, backed by American power, funded by American taxpayers.

As that commitment withdraws, as it has been withdrawing for thirty-five years, the system destabilizes. Regional powers that were constrained by American presence test boundaries. Adversaries that were deterred by American credibility probe for weakness. Allies that depended on American guarantees hedge their bets. What results is exactly what we're seeing: overlapping conflicts, cascading escalation, and a global order that nobody is maintaining because the country that built it is no longer willing or able to do so.

This is World War III. Not the version with mushroom clouds and civilization ending. The version where the global system built after 1945 breaks down, conflict fills the vacuum, and the world reorganizes itself around new power centers through a decade or more of grinding, painful, bloody adjustment. The nuclear arsenals prevent the catastrophic version. They don't prevent the slow version. And the slow version is what we're living through.

. . .

Part Three described the strategic context: a retreating superpower, an obsolete commitment structure, and a world war that's already underway. The temptation is to conclude that the situation is hopeless, that decline is inevitable, that the best anyone can do is manage the descent.

That conclusion would be wrong. The situation is serious, but it's not hopeless. And the reason it's not hopeless has nothing to do with politics, alliances, or military hardware. It has to do with forces that are bigger than any of those things. Forces that will reshape the world more profoundly than any war, any treaty, or any presidential decision.

Demographic collapse. Resource depletion. Climate change. And a fourth horseman that nobody is prepared for.

These forces sound unstoppable. They are not. They are big, slow, and knowable, which means they are addressable by anyone with the tools and the will to act. The tools exist. Whether the will does is the question this book is building toward.

Welcome to Part Four.

123

Chapter 12: The Empty Cradle

The demographic collapse is real and it must be addressed, but I think the framing needs to change. Falling birth rates are partly a function of urbanization. People on farms have large families because children are labor. People in cities do not. That is not going to reverse. So the answer is not to try to force birth rates back up but to redesign our economic and military systems to function with smaller populations. Automation and AI will help bridge the gap in the medium term. Legal immigration is another real answer, and one that gets drowned out in the noise around the illegal version. The countries that figure out how to absorb legal immigrants well and integrate them into the workforce will come out ahead. The ones that close themselves off will shrink.

In 1950, the average woman on Earth had five children. By 2020, that number had fallen to 2.3. By 2024, roughly half the world's countries had fertility rates below the replacement level of 2.1 children per woman. The trend line is pointing toward a world where, sometime in the second half of this century, the global population stops growing and starts shrinking.

That sentence sounds benign. It's not. It's the most consequential demographic shift in human history, and almost nobody understands what it means.

For the entirety of recorded civilization, the assumption has been that there will always be more people tomorrow than today. Every economic system, every pension plan, every military strategy, every government budget is built on the premise of growth. More workers to replace the ones who retire. More taxpayers to fund the obligations the last generation created. More soldiers to fill the ranks. More consumers to buy the products. More young people to take care of the old ones.

What happens when that assumption breaks?

• • •

Start with China, because China is where the demographic crisis is most urgent, most consequential, and most dishonestly reported.

The official Chinese population figures are almost certainly wrong. During the one-child policy era, which ran from 1980 to 2015, local officials had massive incentives to overcount their populations. More people in your district meant more funding from Beijing. Families had incentives to underreport daughters, who were less valued under a patriarchal system that prized male heirs. The combination produced a statistical fog that China has been trying to clear ever since.

Yi Fuxian, a demographer at the University of Wisconsin, has argued for years that China's population is overcounted by roughly 130 million people. His research, based on vaccination records and school enrollment data, suggests that local officials systematically inflated birth figures to secure funding from Beijing. Other demographers dispute the scale of the overcount, but the discrepancies in China's own data are hard to explain away. If the most pessimistic estimates are correct, China's actual population might be closer to 1.28 billion than the officially reported 1.4 billion. That's not a rounding error. That's the population of a major country that doesn't exist.

The working-age population is already shrinking. By 2050, under official projections, China loses over 200 million workers. If the real numbers are worse than official projections, that decline is already further along than anyone in Beijing is admitting publicly. The factories, the construction crews, the service industries, the military, the entire productive apparatus of the world's second-largest economy is running on a workforce that is getting smaller every year.

This is the context for everything China does on the world stage. The Belt and Road Initiative. The push into the Russian Far East. The African investments. The rare earth dominance. The energy transition play. None of it makes sense as the patient strategy of a country with time on its side. It makes sense as the urgent, almost desperate strategy of a country that knows its window is closing.

Xi Jinping knows the numbers. The aggressive posturing of the last decade isn't the confidence of a rising power. It's the urgency of a power that understands it has maybe 15 to 20 years, possibly less, to lock in the structural advantages it needs before the demographic math makes expansion impossible. Everything China is doing right now is a race against its own birth rate.

• • •

China is the most consequential case, but it's not the most extreme.

South Korea's fertility rate hit 0.72 in 2023. That's not below replacement. That's civilizational free-fall. Replacement is 2.1. South Korea is at roughly a third of that. If the rate doesn't change, and there's no evidence it will, the South Korean population halves every generation. A country of 52 million becomes 26 million in thirty years and 13 million in sixty. The math is merciless.

Japan has been in demographic decline since 2008. The population peaked at 128 million and is projected to fall below 100 million by the mid-2050s. Japan has been managing this with characteristic discipline, investing heavily in robotics and automation to compensate for the shrinking workforce. It's the test case for whether a technologically advanced society can maintain its economic output with a declining population. The jury is still out, but Japan even attempting it tells you something about the severity of the problem.

Russia is depopulating faster than any major power. Between the demographic legacy of the Soviet collapse, the exodus of young professionals since the Ukraine invasion, and the catastrophic military casualties of the past four years, Russia is losing population at a rate that threatens its ability to function as a state. The Russian Far East, with fewer than 8 million people spread across a territory larger than most continents, and that number shrinking by the year, is the starkest example of demographic vacancy on the planet. The population has fallen by more than 10 percent in the past two decades despite repeated Kremlin development initiatives. That vacancy is China's opportunity, as we'll see in Part Five.

Europe is aging into irrelevance. Italy's fertility rate is 1.20. Spain's is 1.12. Germany's is 1.36. The European Union as a whole is well below replacement, and the population is graying rapidly. By 2050, the median age in several European countries will be over 50. These are not societies that project military power or drive global economic growth. They're societies that manage retirement systems.

· · ·

The United States is the exception that should be a lesson.

American fertility has fallen to about 1.6, below replacement but not catastrophically so. More importantly, the US has something almost no other developed nation possesses: a functioning immigration system and a cultural identity that, however imperfectly and contentiously, absorbs immigrants and turns them into Americans.

Immigration is the US demographic safety valve. It's why the American population is still growing while Japan's shrinks and Europe's ages out. Every year, roughly a million legal immigrants plus an uncertain number of undocumented arrivals add to the American population. These immigrants are disproportionately young, disproportionately working-age, and disproportionately motivated. They start businesses at higher rates than native-born Americans. They fill labor gaps from agriculture to technology. They have children at rates that help offset declining native fertility.

This is not a cultural argument. This is a national survival argument. In a world where every major competitor faces demographic collapse, the country that maintains its working-age population through immigration has a structural advantage that compounds over decades. The US has that advantage. And it's politically trying to weld the valve shut.

The immigration debate in America is conducted almost entirely in cultural and security terms: who gets in, what they look like, whether they'll learn English, whether they'll take jobs. Almost nobody frames it in demographic and strategic terms: without immigration, the US ages like Europe, its workforce shrinks, its tax base erodes, its military can't recruit, its innovation engine slows, and its competitive position against China deteriorates. Immigration isn't a policy choice. It's a survival strategy. And the people fighting it hardest are the ones most likely to doom their own country.

The numbers tell the story. The National Foundation for American Policy found that immigrants or their children founded 319 of America's 582 billion-dollar startups, 55 percent, with a collective valuation of $1.2 trillion. India alone produced the founders of 66 of those companies. Israel produced 54. The United Kingdom 27. These are not people taking jobs from Americans. They are people creating jobs for Americans, building companies that employ hundreds of thousands of workers, pay billions in taxes, and produce the technologies that maintain the country's competitive edge. Every one of them came to the United States because the United States was the best place on Earth to build something. If that stops being true, they go somewhere else. And wherever they go, the companies go with them.

The STEM pipeline is even more stark. In American graduate programs in engineering, computer science, and mathematics, foreign students routinely make up 50 to 70 percent of enrollment. They earn their degrees at American universities, using American labs, funded by American research grants, and then many of them leave, because the immigration system makes it easier to go home or move to Canada or Australia than to stay in the country that trained them. The H-1B visa lottery, the years-long green card backlog for Indian and Chinese nationals, the lack of a startup visa, the hostile rhetoric that signals they are not welcome: all of it pushes talent out the door that American universities spent years and billions of dollars developing. This is not an immigration policy. It is a talent export program disguised as one.

• • •

The demographic crisis interacts with every other force described here.

Military readiness depends on a population young enough to serve. The US military is already struggling to meet recruitment targets. Expanding the fleet, building the ships, staffing the factories, as described in Part One, requires workers who don't exist in sufficient numbers unless immigration continues.

Economic output depends on a workforce large enough to produce goods and services. The pension and social security systems that every developed nation depends on were designed for growing populations. When the ratio of workers to retirees inverts, the math breaks. Either benefits get cut, taxes go up, or the system goes bankrupt. Most countries will experience some combination of all three.

Doing the math: simple and merciless. When Social Security was created in 1935, there were roughly 40 workers paying into the system for every retiree drawing from it. By 1960, the ratio was 5 to 1. Today it is about 2.8 to 1. By 2035, it will be 2.3 to 1. The trust fund is projected to be depleted by the early 2030s, after which incoming payroll taxes will cover only about 80 percent of scheduled benefits. Nobody in Washington wants to talk about this because every solution is politically painful: raise the retirement age, cut benefits, increase payroll taxes, or some combination. The demographic math makes all of these worse every year because the numerator, workers, is shrinking relative to the denominator, retirees, and the denominator is living longer. An 80-year-old in 2026 consumes far more in healthcare than an 80-year-old in 1970, and there are far more of them.

Japan shows where this leads. The country now spends over a third of its government budget on social security and healthcare for the elderly. Its national debt exceeds 260 percent of GDP, the highest in the developed world, funded primarily by domestic savings that are themselves shrinking as retirees draw down what they saved during their working years. Japan has managed this through cultural discipline, low interest rates, and a willingness to accept economic stagnation that no other democracy has been tested on. Three decades of zero growth have not produced a political revolution in Japan because the Japanese public accepted the trade-off. It is far from clear that the American public would accept the same.

Innovation depends on a deep talent pool. The breakthroughs that drive technological progress, the ones that might solve the other crises in this book, come from a tiny fraction of the population with exceptional ability. Shrink the population and you shrink that fraction in absolute terms. Restrict immigration and you exclude the foreign-born talent that has driven a disproportionate share of American innovation: immigrants or their children founded more than half of America's billion-dollar startups.

And geopolitical competition increasingly comes down to who has the people. Not just to fill armies, though that matters. To fill factories, labs, shipyards, and classrooms. To build the things that need building. To maintain the systems that keep a modern civilization running. A country that runs out of people runs out of options.

• • •

The optimistic case is real, and it has two parts.

First, technology. If AI and automation can compensate for shrinking workforces, as Japan is attempting, then the link between population size and economic output can be weakened or broken. Chapter 23 will explore this in detail. If longevity science extends productive lifespans, as Chapter 24 will argue, then the dependency ratio improves even without new births: a 70-year-old with the function of a 40-year-old is a worker, not a retiree. Technology doesn't eliminate the demographic challenge, but it can change its shape from a cliff to a slope.

Second, immigration. For the United States specifically, the solution is already available and already working. The country just has to keep doing what it's been doing for 250 years: welcoming people who want to come here and giving them a reason to stay. That's it. That's the entire demographic strategy. It's not complicated. It's just politically contested by people who don't understand what's at stake.

The pessimistic case is equally real. Not every country has the US immigration advantage. China can't absorb immigrants at scale because its culture and governance don't support it. Japan has tried modest immigration reforms and found them culturally wrenching. Russia's demographic decline is being accelerated by war and emigration at the same time. For these countries, the workforce cliff is real, it's approaching, and there's no obvious fix short of the technological breakthroughs that may or may not arrive in time.

Demographics is destiny. It always has been. The civilizations that thrived had growing, healthy, educated populations. The ones that declined ran out of people, or ran out of people willing to do the work that needed doing.

The 21st century will be defined by which countries manage the demographic transition and which ones don't. The tools to manage it exist. The question, as always, is whether the political will exists to use them.

But demographics is only the first horseman. The second one connects to something most people never think about when they sit down to eat dinner. Where their food comes from. What it's made of. And what happens when the chemical that makes half the world's food supply possible starts running out.

But first, we need to talk about rivers. And why a continent with 1.4 billion people and the youngest population on Earth still can't catch up.

Chapter 13: The Continent That Can't Connect

There's a reason the United States became an economic superpower, and it's not the one most Americans learn in school. It wasn't democracy, though that helped. It wasn't entrepreneurial spirit, though there was plenty. It wasn't the Protestant work ethic or manifest destiny or the wisdom of the Founding Fathers.

It was a river.

The Mississippi-Missouri river system is the largest navigable waterway network on Earth. Over 12,000 miles of commercially navigable rivers draining the most productive agricultural land on the planet, connecting the deep interior of the continent to the Gulf of Mexico and from there to the world's oceans. Barges on the Mississippi can move bulk commodities, grain, coal, steel, chemicals, at roughly one-tenth the cost of rail and one-thirtieth the cost of trucking. The system was there before the first European settler arrived. Nature built it. The United States just had to show up and use it.

Cheap bulk transport from the interior to the coast is the foundation of an integrated economy at continental scale. It's the reason the American Midwest could become the breadbasket of the world. It's the reason heavy industry could locate in Pittsburgh and Chicago and St. Louis rather than clustering at port cities. It's the reason the United States could build a national market before the railroad era and then turbocharge that market once rail arrived. The Mississippi system didn't just help the American economy. It made the American economy possible.

When analysts talk about Africa's demographic potential, about the 2.5 billion people the continent will hold by 2050 and the "demographic dividend" that could power the 21st century's economic growth, they rarely talk about rivers. They should. Because Africa doesn't have a Mississippi. And that single geographic fact may matter more than all the investment, all the aid, and all the optimistic projections combined.

. . .

Look at a map of African rivers and compare it to a map of American rivers. The difference tells you most of what you need to know.

The Congo River is the second-largest in the world by discharge, surpassed only by the Amazon. It drains an enormous basin in central Africa. On paper, it should be Africa's Mississippi. In practice, it's largely unnavigable for commercial traffic. A series of cataracts and rapids, including the famous Inga Falls, break the river into segments that can't be traversed by cargo vessels. The lower Congo drops 900 feet in 200 miles, making it one of the most turbulent stretches of river on the planet. You can float a canoe through parts of it. You can't push a barge loaded with grain.

The Niger River is West Africa's major waterway. It takes a bizarre boomerang course, flowing northeast from Guinea into the Sahel before looping back south through Nigeria to the Gulf of Guinea. It serves some commercial traffic in limited stretches, but it doesn't connect the productive agricultural interior to the coast in a way that enables bulk trade. The seasonal variation in water level, from flood to near-dry, makes year-round navigation impossible in many sections.

The Nile is the one African river everyone knows, and it serves exactly one narrow corridor. Egypt's economy has depended on the Nile for five thousand years, but the Nile doesn't connect East Africa to North Africa in a commercially useful way. The cataracts in Sudan break the navigable stretches. The seasonal flooding that once made Egyptian agriculture possible is now controlled by the Aswan Dam, which solved one problem and created others.

The Zambezi has Victoria Falls in the middle of it. The Orange River in South Africa is too shallow for large-vessel navigation. The Limpopo is seasonal. None of Africa's rivers provide what the Mississippi provides: a continuous, year-round, navigable highway from the interior to the sea.

This isn't a failure of African development. It's a fact of African geology. The continent is a high plateau ringed by coastal lowlands, and the rivers that descend from the plateau to the coast drop through rapids, cataracts, and gorges that make navigation impossible. The Mississippi flows gently across a vast flat plain. African rivers fall off cliffs.

• • •

Why does this matter for the 21st century? Because moving heavy things cheaply is the foundation of industrial civilization, and Africa can't do it.

When there are no navigable rivers, you need roads and railroads. Africa has some of both, mostly built during the colonial era and mostly designed to move extracted resources from mines to ports rather than to connect African markets to each other. The colonial powers built rail lines that ran from the interior to the coast, from copper mine to shipping terminal, from diamond field to harbor. They did not build networks that connected African cities to one another, because connecting African cities to one another didn't serve the extraction model.

After independence, most of those rail systems deteriorated. Maintenance requires funding that cash-strapped governments couldn't provide. Roads were built but potholes appeared faster than repairs. The trans-African highway network, on paper, connects the continent. In practice, many stretches are impassable during the rainy season, potholed into ruin during the dry season, or simply not maintained at all. A truck carrying goods from Lagos to Nairobi faces a journey of weeks through border crossings, road conditions, and bureaucratic obstacles that make the cost of overland transport astronomically higher than it would be on a navigable river.

What results is that African economies are isolated from one another. Intra-African trade accounts for only about 15 percent of total African trade. Compare that to intra-European trade, which is over 60 percent, or intra-Asian trade at over 50 percent. African countries trade with Europe and China far more than they trade with each other, because it's literally cheaper to ship goods across an ocean than to truck them across the continent.

The African Continental Free Trade Area, or AfCFTA, launched in 2021, is the most ambitious attempt to change this. It aims to create a single continental market for goods and services, covering 1.4 billion people with a combined GDP of over \$3 trillion. On paper, it is the largest free trade area in the world by number of participating countries.

In practice, it is a framework agreement that has yet to produce meaningful trade liberalization. Tariff negotiations are incomplete. Rules of origin are unresolved. The customs infrastructure needed to process cross-border trade efficiently does not exist at most African border crossings. And the physical infrastructure, the roads, rails, and bridges that would carry the goods, is what this chapter has been describing: inadequate, fragmented, and designed to move resources from interior to coast rather than from country to country.

The AfCFTA is the right idea. It is an idea that will take decades to implement, and implementation requires the very infrastructure investments that no African government has the fiscal capacity to make on its own. This is why Chinese investment matters so much, and why the terms of that investment matter more than the investment itself. If Chinese-built infrastructure connects Africa to China, it deepens dependency. If it connects Africa to itself, it builds sovereignty. So far, most of it has done the former.

A demographic dividend requires an integrated economy. 2.5 billion people scattered across isolated markets with poor transportation links between them is not a dividend. It's 2.5 billion people in 54 countries that can't efficiently trade with each other.

• • •

China is trying to fix this, and it's worth understanding what Chinese investment in Africa means in geographic terms.

The Belt and Road Initiative has funded railroads, ports, highways, and bridges across the continent. The Addis Ababa-Djibouti Railway connects the Ethiopian capital to the coast. The Mombasa-Nairobi Standard Gauge Railway, built by the China Road and Bridge Corporation at a cost of roughly $5 billion, 90 percent financed by China's Export-Import Bank, modernized a critical Kenyan corridor when it opened in 2017. But Kenya now spends over $1 billion a year servicing the debt, the railway was supposed to extend to Uganda and never did because China stopped funding it 75 miles short of the border, and the IMF has classified Kenya as being at high risk of debt distress. The railway connects Kenya's port to Kenya's capital. It does not connect Kenya to its neighbors. That distinction matters. Chinese-built ports in Djibouti, Lamu, Dar es Salaam, and elsewhere are expanding Africa's connection to global shipping lanes. These are real infrastructure projects that are moving real goods and creating real economic connections that didn't exist before.

But here's the catch. You can't engineer your way around fundamental geography, and the cost of trying is staggering.

The Mississippi system cost the United States nothing to build. Nature provided it for free. Every mile of African railroad costs money to construct and money to maintain, through terrain that ranges from desert to jungle to mountain. A single rail line from the interior to the coast serves the customers along that line. It doesn't create the continental-scale network that the Mississippi provides, because a network requires thousands of miles of interconnected routes, and each mile has to be planned, funded, built, and maintained in perpetuity.

The maintenance problem is the one that kills optimistic projections. Chinese loans fund construction, but maintenance has to be funded by the operating country, and African governments generally don't have the revenue base to maintain infrastructure at the level required for it to function reliably over decades. Railroads deteriorate. Roads develop potholes. Ports need dredging. Equipment breaks down and replacement parts have to be imported. The colonial-era infrastructure deteriorated for exactly this reason, and there's no structural change that prevents the same thing from happening to Chinese-built infrastructure.

Chinese investment in Africa buys influence and creates resource extraction corridors. It does not create the integrated continental economy that would turn Africa's demographic weight into economic power. The infrastructure follows the extraction model, mine to port, resource to ship, just as the colonial infrastructure did. It connects Africa to China, not Africa to itself.

• • •

Then layer climate change on top of the geographic constraints.

Africa is the continent most vulnerable to climate disruption, and the disruption is hitting the places where the population is growing fastest.

The Sahel, the semi-arid band stretching across sub-Saharan Africa from Senegal to Sudan, is expanding southward as the Sahara grows. Agricultural land that supported communities for generations is turning to desert. Lake Chad, which was one of the largest lakes in Africa and the water source for tens of millions of people, has shrunk by roughly 90 percent since the 1960s. The communities that depended on it are migrating, fighting, or dying.

Coastal West Africa, home to megacities like Lagos, Accra, and Abidjan, faces sea-level rise that threatens to inundate low-lying areas where millions live. Lagos, a city of over 20 million people built largely on wetlands and lagoons at sea level, is one of the most climate-vulnerable cities on Earth. The infrastructure to protect it, sea walls, drainage systems, elevated construction, would cost tens of billions of dollars that Nigeria doesn't have.

East Africa faces a different pattern: more intense rainfall and flooding in some areas, prolonged drought in others, and the retreat of glaciers on Mount Kenya and Kilimanjaro that feed rivers supporting millions. The agricultural systems that most East Africans depend on are calibrated to historical rainfall patterns that no longer hold.

Southern Africa is drying. Cape Town nearly ran out of water in 2018. The region's agricultural productivity is projected to decline significantly as temperatures rise and rainfall becomes more erratic.

The population is booming in exactly the places that are becoming less habitable. Nigeria, which is projected to become the third-most-populous country on Earth by 2050, is at the same time losing agricultural land to desertification in the north and coastal land to flooding in the south. The Democratic Republic of Congo, which has the second-largest tropical rainforest on Earth and some of the richest mineral deposits, is also losing forest cover at an accelerating rate, degrading the ecosystem that regulates the regional climate.

• • •

So what does the African demographic story mean for the geopolitical competition?

The optimistic narrative says Africa is the demographic prize. The only continent with explosive youth population growth. The labor force of the future. Whoever wins Africa wins the century.

The realistic assessment is more complicated, and it starts with honesty about geography.

Africa's 2.5 billion people in 2050 are a potential asset of world-historical significance. Young, growing, hungry for opportunity. But potential is not destiny. Potential requires infrastructure to connect markets, governance to maintain that infrastructure, education to develop the workforce, and a climate stable enough to grow the food that keeps 2.5 billion people alive.

Without navigable rivers, the infrastructure has to be built from scratch, at enormous cost, and maintained forever. Without stable governance, the infrastructure deteriorates as fast as it's built. Without education, the demographic mass doesn't translate into productive labor. Without climate stability, the agricultural base that feeds the population erodes underneath it.

The demographic dividend is not automatic. It never was. It requires specific conditions that Africa mostly doesn't have and that are getting harder to create as the climate shifts. China's investment is real but follows an extraction model that serves Chinese interests, not African development. The Western aid model has a half-century track record of failure. The African Union's own development plans are ambitious on paper and underfunded in practice.

None of this means Africa is doomed. It means that anyone who tells you Africa will save the 21st century without addressing rivers, roads, climate, and governance is selling you something. The continent's people are its greatest resource. Turning that resource into economic power requires solving geographic and infrastructure challenges that have defeated every previous attempt, with climate change making the problem harder every year.

The demographic story of the 21st century isn't just about how many people there are. It's about where they are, what connects them, and whether the land they're standing on will still feed them in thirty years. For Africa, the answers to those questions are uncertain at best and alarming at worst.

And the land question hides a deeper one. Because the thing that feeds half the world isn't land or water or sunlight. It's a chemical process that runs on natural gas. And natural gas is running out.

Chapter 14: The Last Barrel and the Empty Plate

In 1909, a German chemist named Fritz Haber figured out how to pull nitrogen out of the air and combine it with hydrogen to make ammonia. His colleague Carl Bosch figured out how to do it at industrial scale. The Haber-Bosch process, as it became known, is arguably the most important invention in human history. More important than the printing press. More important than the steam engine. More important than the computer.

That sounds like an absurd claim until you understand what it did. Before Haber-Bosch, the amount of food the Earth could produce was limited by the amount of nitrogen available in the soil. Nitrogen is the essential nutrient for plant growth. Without it, crops don't grow. The natural sources of nitrogen, manure, crop rotation, guano deposits on remote Pacific islands, could support a global population of maybe 3 to 4 billion people. That was the ceiling. Humanity was approaching it, and famine was the historical mechanism that enforced it.

Haber-Bosch shattered the ceiling. Synthetic nitrogen fertilizer allowed farmers to grow far more food per acre than natural soil fertility could support. The Green Revolution of the mid-20th century, which is credited with saving a billion people from starvation, was built on synthetic fertilizer. The global population went from 2 billion in 1927 to 8 billion in 2022. The difference, the roughly 4 billion people alive today who wouldn't be alive without synthetic nitrogen, is the Haber-Bosch dividend.

Half the nitrogen atoms in your body passed through the Haber-Bosch process. Half the food you eat exists because of it. It is, without exaggeration, the chemical reaction that sustains human civilization at its current scale.

And it runs on natural gas.

. . .

The Haber-Bosch process requires two inputs: nitrogen from the atmosphere, which is free and unlimited, and hydrogen, which has to be produced from something. The cheapest and most efficient source of hydrogen is natural gas, specifically methane. You take methane, react it with steam at high temperature and pressure, and get hydrogen and carbon dioxide. Then you combine the hydrogen with nitrogen under even higher temperature and pressure, and you get ammonia. The ammonia becomes the feedstock for every nitrogen fertilizer product on the market.

Roughly 3 to 5 percent of the world's natural gas production goes to making ammonia. That doesn't sound like a lot until you realize that the ammonia makes roughly half the world's food. Take away the natural gas, and you don't just lose a fuel source. You lose the chemical that feeds 4 billion people.

This is the connection that almost nobody makes when they talk about energy transitions and oil depletion. When environmentalists talk about getting off fossil fuels, they're thinking about cars and power plants and heating systems. Those are solvable problems. You can run a car on electricity. You can heat a building with a heat pump. You can generate electricity from solar and wind and nuclear.

You cannot run Haber-Bosch on intermittent power. Not at the scale needed to feed billions. The process requires sustained high temperature and high pressure, which means sustained, reliable energy input. A cloud passing over a solar farm or the wind dying down doesn't just reduce your electricity output. It interrupts a chemical reaction that has to run continuously to be efficient. You can't turn ammonia synthesis on and off like a light switch.

And natural gas isn't just the energy source for Haber-Bosch. It's the chemical feedstock. The hydrogen that goes into the ammonia comes from the methane molecule itself. Replacing the energy is one problem. Replacing the chemistry is a different, harder problem.

• • •

Natural gas is not running out tomorrow. Let's be clear about that. This is not a Zeihan-style prediction that the wells go dry next Tuesday and civilization collapses. Global natural gas reserves are substantial, and new extraction technologies continue to find more. The timeline for serious depletion is measured in decades, not years.

But decades is exactly the planning horizon this book operates on. And on that horizon, the trajectory is clear: the easy gas is gone. Each new well is more expensive to drill, more technically demanding to operate, and produces less output per dollar invested than the ones before it. The cost curve is rising. It has been rising for years. And at some point, the cost of natural gas rises enough that the cost of fertilizer follows it, and the cost of food follows that.

This isn't a cliff. It's a slope. But a slope in the cost of something that half the world depends on for survival is a slope that leads to catastrophe for the people at the bottom of the economic ladder. Rich countries can absorb higher food prices. They grumble and adjust budgets and eat slightly less meat. Poor countries can't. When food prices spike, people in the developing world don't adjust their budgets. They starve. Or they migrate. Or they fight.

The 2008 food price crisis triggered riots in over 30 countries. The 2011 food price spike was one of the contributing factors to the Arab Spring, which toppled governments across the Middle East and North Africa and led to the Syrian civil war, which produced a refugee crisis that destabilized European politics for a decade. Food price shocks don't stay in the grocery store. They cascade through political systems with a speed and violence that surprises people who've never been hungry.

• • •

But nitrogen is only one leg of the fertilizer stool. The other two are phosphorus and potassium, and their stories are if anything scarier because the solutions are less obvious.

Phosphorus comes from phosphate rock, which is mined. The world's phosphate reserves are concentrated in a handful of countries. Morocco and Western Sahara hold roughly 70 percent of known global reserves. China, which restricts exports to protect its own agricultural supply, holds a distant second. The United States, which was once a major phosphate producer, has largely depleted its high-quality deposits.

Phosphate rock is a finite resource. Unlike nitrogen, which is literally pulled from the atmosphere, phosphorus has to be dug out of the ground. There is no atmospheric phosphorus to synthesize. When the mines run out, they're out. Estimates of how long the reserves will last vary widely, from 50 years to several centuries, depending on assumptions about consumption rates, undiscovered deposits, and the economic viability of lower-grade ores. But the direction is not in dispute: the easiest, cheapest, highest-quality deposits are being mined first, and what remains will cost more to extract and process.

Morocco's dominance of the phosphate market should set off the same alarm bells that Middle Eastern oil dominance set off fifty years ago. A single country controlling the critical input for global food production is a strategic vulnerability of the first order. If Morocco experienced political instability, or if it decided to use its phosphate reserves as a geopolitical weapon the way OPEC used oil in 1973, the consequences for global agriculture would be immediate and severe.

Potassium, the third essential nutrient, comes from potash deposits. Canada, Russia, and Belarus are the major producers. Canada's reserves are enormous, which is strategically convenient for the United States. But Russia and Belarus together account for much of global production, and sanctions related to the Ukraine war have already disrupted potash supply chains and contributed to fertilizer price spikes.

Three essential nutrients. Three finite supply chains. Three potential chokepoints. The entire global food system depends on all three arriving reliably and affordably at farms around the world, and any disruption to any of them translates directly into hunger.

• • •

The Iran war demonstrated this in real time, and most people didn't notice.

Roughly a third of the world's fertilizer trade transits the Strait of Hormuz. When Iran closed the strait in early March 2026, fertilizer shipments stopped along with everything else. Urea prices, the benchmark for nitrogen fertilizer, jumped from $475 per metric ton to $680 within days. This happened during the spring planting season in the American Midwest, when farmers were buying fertilizer for corn and soybean planting.

The United States is the world's largest food exporter. Corn and soybeans from the Midwest feed livestock and people across the planet. When the fertilizer to grow that corn and those soybeans gets more expensive, or doesn't arrive at all, the consequences ripple through the global food supply within months. It's not a hypothetical. It happened. Fertilizer prices spiked, planting decisions changed, and the downstream effects on food prices were already emerging before the war's first month was over.

And this was a temporary disruption. The strait will reopen eventually. The fertilizer will flow again. Prices will come back down. But what the episode demonstrated is how fragile the system is. A single shipping chokepoint, disrupted for a few weeks, can send food prices spiking across the planet during the one window of the year when the planting decisions that determine the next harvest are being made.

Now project that fragility forward. Not a temporary disruption from a war, but a permanent, structural increase in the cost of natural gas, and therefore ammonia, and therefore nitrogen fertilizer, as global gas production passes its peak and begins to decline. Not a spike that goes back down. A slope that only goes up.

• • •

The chain runs like this, and it's worth tracing the whole thing because most people have never seen it laid out:

Natural gas comes out of the ground. It goes to an ammonia plant. The plant uses it as both fuel and chemical feedstock to produce ammonia. The ammonia goes to a fertilizer factory that combines it with other chemicals to make urea, ammonium nitrate, or other nitrogen fertilizers. The fertilizer gets shipped to a distribution center. From there it goes to a farm supply dealer. The farmer buys it and applies it to fields. The crops grow. The crops feed livestock or go directly to food processing. The food gets shipped to grocery stores. You eat it.

Every step in that chain adds cost. Every step depends on transportation, which depends on fuel, which depends on the same fossil fuels that started the chain. The truck that delivers the fertilizer runs on diesel. The tractor that applies it runs on diesel. The combine that harvests the crop runs on diesel. The train that moves the grain to the processing plant runs on diesel. The ship that exports it runs on heavy fuel oil. The entire food system, from field to fork, is saturated with fossil fuel at every stage.

When people talk about the energy transition, they rarely account for this. Electrifying cars is relatively simple. Electrifying the global food supply chain, from gas well to dinner plate, is a challenge of a completely different magnitude. And it starts with the hardest part: finding a way to make ammonia without natural gas.

• • •

The solutions exist. They're real. They're not science fiction. But they're not deployed at scale, and getting them to scale is the engineering challenge that will determine whether 10 billion people eat in 2060 or whether 4 billion of them face famine.

Nuclear-powered hydrogen production can replace natural gas as the feedstock for ammonia synthesis. You use nuclear heat and electricity to split water molecules through electrolysis, producing hydrogen and oxygen. The hydrogen goes into the Haber-Bosch process the same way methane-derived hydrogen does. The technology works. It's been demonstrated. What it requires is nuclear power plants, which the world has spent fifty years refusing to build at scale because of irrational fear driven by three accidents across seven decades of operation. We'll come back to this in Chapter 22.

AI-driven precision agriculture can reduce the amount of fertilizer needed per unit of food produced by 30 to 50 percent while maintaining or increasing yields. Chapter 23 will explain how. For now, the point is that the technology exists to break the linear relationship between natural gas consumption and food production. But it requires AI, sensors, and equipment that most of the world's farmers do not have. Deployment at global scale does not exist. Yet.

Genetic engineering could crack the problem entirely. Some plants, primarily legumes like soybeans and clover, host bacteria in their root nodules that can fix atmospheric nitrogen directly, no Haber-Bosch required. If genetic engineering can give this capability to cereal crops, rice, wheat, and corn, the dependency on synthetic nitrogen fertilizer breaks. AI-powered molecular modeling and protein engineering tools are making this kind of biological moonshot more achievable than at any previous point in history. It's not guaranteed. It's not imminent. But it's plausible on a 20-to-30-year timeline, which is the timeline that matters.

Each of these solutions connects to other chapters. Nuclear energy in Chapter 22. AI in Chapter 23. Genetic engineering in Chapter 25. The tools are real. The question, as always, is whether they'll be deployed with strategic intent or whether we'll stumble toward the cliff arguing about who should pay for the bridge.

• • •

Here's what the oil-fertilizer-food chain means for the geopolitical model.

The countries that control fertilizer inputs, natural gas, phosphate, and potash, hold strategic leverage that will increase every year as those resources become scarcer and more expensive. Russia is a major fertilizer exporter. China controls significant phosphate reserves and restricts exports.

Canada has massive potash deposits, which is strategically fortunate for the United States. Morocco's phosphate monopoly makes it one of the most strategically important countries on the planet, though almost nobody in Washington treats it that way.

The countries that can produce food at scale, the US, Brazil, Canada, Australia, Ukraine (when it's not being invaded), hold a different kind of leverage: the ability to feed or starve other nations. In a world of rising food prices and declining fertilizer availability, food exporters become as strategically important as oil exporters were in the 20th century.

The United States holds the strongest hand. Enormous natural gas reserves to feed its own ammonia production. Canadian potash next door. The most productive agricultural land in the world, connected by the Mississippi system to global export markets. If the US manages its fertilizer security the way it should manage its energy security, decoupling from global commodity price shocks through strategic reserves and domestic supply priority, it could be the most food-secure nation on Earth for the rest of this century.

But that's an *if*. And right now, the US isn't managing its food security strategically. It's treating agriculture as a commodity business and fertilizer as a market good, subject to the same price volatility and supply disruptions as everything else that flows through global chokepoints. The Hormuz closure proved how fast that system breaks. The structural depletion of natural gas will prove it again, slower but harder to fix.

Oil depletion isn't just an energy crisis. It's a food crisis. The food crisis is a migration crisis. The migration crisis is a political crisis. The chain runs from a gas wellhead in Texas to a famine in sub-Saharan Africa to a border wall in Arizona, and almost nobody in the political conversation traces the full connection.

This book does. And the connection gets worse, because the physical geography that the food chain sits on is moving. The map is not fixed. It never was. But the rate at which it is changing now is something no civilization has experienced.

Chapter 15: The Burning Map

Every strategic analysis here, every chapter about fleet sizes and missile stocks and alliance structures and trade routes, sits on top of a physical map. Geography. Where the mountains are. Where the rivers flow. Where the coasts run. Where the arable land lies. Where the water is. Geography is the foundation that everything else is built on, and for the entirety of human strategic thinking, that foundation has been treated as fixed.

It's not fixed anymore. The map is moving. And when the map moves, everything built on top of it shifts.

. . .

This chapter is not a moral argument about climate change. If you want guilt and polar bears and carbon footprint calculators, there are plenty of books for that. This chapter is about what happens to the physical geography that nations depend on when temperatures rise, rainfall patterns shift, sea levels climb, and the places where people grow food and build cities stop being the places where food grows and cities survive.

The geopolitical implications are larger than most climate books acknowledge, because most climate books are written by environmental scientists, not strategists. And the strategic implications are larger than most geopolitics books acknowledge, because most geopolitics books were written before the map started moving fast enough to matter within a single generation.

This book sits at the intersection. And the intersection is where it gets dangerous.

. . .

Start with the winners, because they exist and because their existence changes everything about the power competition described in the rest of this book.

Northern latitudes get warmer. That's bad if you're a polar bear, but it's very good if you're a farmer in Canada or Scandinavia or the Russian Far East. Warmer temperatures mean longer growing seasons. Longer growing seasons mean more food from the same land, or food from land that was previously too cold to farm at all. The permafrost line in Siberia is retreating northward, exposing millions of acres that have never been cultivated. Canada's prairies are expanding their productive window. Scandinavia is growing crops that were impossible a generation ago.

This doesn't mean northern countries will become tropical paradises. The soils under melting permafrost are often poor. The infrastructure to farm newly available land doesn't exist. The transition from frozen tundra to productive farmland takes decades, not years. But the direction is clear: the agricultural map is shifting northward, and the countries that own northern land are gaining an asset that appreciates in value every year.

The Russian Far East, the territory that Chapter 19 argues China is positioning to absorb, becomes more valuable with every degree of warming. It's not just the minerals and the timber and the freshwater. It's the agricultural potential of a warming Siberia. China's play for the Far East isn't just about today's resources. It's about tomorrow's farmland. And on a planet where the places that currently grow food are losing that capacity, tomorrow's farmland may be the most valuable strategic asset on Earth.

Canada becomes more important than most people realize. Already the world's second-largest country by area, already a major agricultural producer, Canada gains arable land as the climate warms. Its freshwater reserves are among the largest on the planet. Its location, sharing the world's longest undefended border with the United States, makes it a natural partner in a hemispheric model. The US-Canada combination, in a warming world, holds an agricultural and water advantage that no other power block can match.

• • •

Now the losers. And the list is long.

The tropics and subtropics, where the majority of the world's population lives, face a convergence of heat, water stress, and sea-level rise that threatens to make large areas functionally uninhabitable within the lifetimes of people already born.

South and Southeast Asia face the most dramatic impacts. Bangladesh, a country of 170 million people, is one of the most flood-prone nations on Earth. Much of it sits less than five meters above sea level. A one-meter rise in sea level, which is within the range of mainstream projections for this century, would displace tens of millions of people. Where do they go? India, which has its own climate problems, including groundwater depletion, heat waves that are already killing thousands, and agricultural disruption across its northern breadbasket.

Vietnam's Mekong Delta, which produces roughly half the country's rice, is sinking and flooding at the same time. Indonesia's capital, Jakarta, is sinking so fast that the government has decided to build an entirely new capital on Borneo. The Philippines faces increasingly destructive typhoons. These aren't future projections. They're happening now.

The Middle East and North Africa are drying out. Iran has been dealing with catastrophic water shortages for years. Lake Urmia, once one of the largest salt lakes in the world, has

154

nearly disappeared. The agricultural regions that feed Iran's 85 million people are collapsing. Tehran, a city of 9 million, has been discussing relocating the capital because the water situation is becoming untenable. And then a war arrived on top of the water crisis. Climate didn't cause the Iran war, but it made Iran weaker going in and will make recovery harder coming out.

Iraq's Tigris and Euphrates rivers are declining as Turkey builds dams upstream and rainfall decreases. Jordan is one of the most water-stressed countries on the planet. Syria's civil war was preceded by the worst drought in the region's recorded history, which drove farmers off the land and into cities where they found no jobs and no future. The connection between drought and revolution isn't theoretical. It's documented.

The Mediterranean basin, which includes southern Europe, is on a drying trajectory that will transform the region over the coming decades. Spain and Italy are experiencing water stress that is already affecting agriculture. Greece is burning. Portugal is burning. The tourism economies that sustain much of southern Europe depend on a climate that is becoming less pleasant and more dangerous every year.

• • •

Water is the new oil. That phrase has become a cliché, but clichés become clichés because they're true.

Freshwater scarcity is already driving conflict. The Nile Basin is a tension zone where Egypt, Sudan, and Ethiopia are in a slow-motion confrontation over the Grand Ethiopian Renaissance Dam, which controls flow to downstream nations that depend on the river for survival. Egypt has said, repeatedly and at the highest levels, that the dam is an existential threat. That's not diplomatic posturing. Egypt is a desert with a river running through it. Without the Nile, Egypt ceases to function.

The Tigris-Euphrates system, as mentioned, is declining. The Indus River, which feeds Pakistan's agriculture and provides water for over 200 million people, depends on glacial meltwater from the Himalayas. As the glaciers retreat, the river's flow will eventually diminish. In the near term, accelerated melting increases flow and flooding. In the long term, the source runs dry. Pakistan's water future is a ticking clock.

Lake Baikal in Siberia holds roughly 20 percent of the world's surface freshwater. It's another reason China wants the Russian Far East. Not just minerals and farmland and Arctic access. Water. On a planet that's running short of it, controlling a fifth of the world's freshwater supply is a strategic asset beyond calculation.

The countries that have abundant freshwater, Canada, Russia, Brazil, and to a lesser extent the United States, hold strategic positions that will appreciate with every degree of warming. The countries that don't, which includes most of the Middle East, North Africa, Central Asia, and increasingly parts of South Asia and southern Europe, face a future of rationing, conflict, and decline.

The worst-case example of what happens when a society destroys its own water supply is the Aral Sea, and most Americans have never heard of it. In the 1960s, the Soviet government diverted the rivers feeding the Aral Sea to irrigate cotton fields across Central Asia. The fourth-largest lake on Earth began to shrink. By 2020, it had lost roughly 90 percent of its volume. Fishing villages that once sat on the shore now stand a hundred kilometers from the receding waterline. The exposed lakebed is a salt flat laced with decades of pesticide residue from the cotton fields, and the dust blows across Uzbekistan and Kazakhstan causing respiratory disease, anemia, and cancer clusters in communities downwind. Satellite photographs from 1960 next to images from today show a body of water the size of Ireland reduced to a few toxic ponds. It is the largest man-made environmental disaster in history, and it was caused by an irrigation policy.

China has 20 percent of the world's population and 7 percent of its freshwater. The imbalance is structural and unfixable by any means short of moving the population or moving the water. Beijing chose to move the water. The South-to-North Water Transfer Project, the largest water infrastructure project in human history, has spent over $80 billion building canals that divert water from the Yangtze River basin in the south to the parched North China Plain, where 40 percent of the country's population lives, most of its wheat grows, and the capital sits on an aquifer that is depleting so fast the city is physically sinking. The eastern and central canal routes function. The western route, which would tap rivers on the Tibetan Plateau and solve the long-term problem, has not been built because the engineering through some of the most rugged terrain on Earth is borderline impossible.

Meanwhile, the groundwater underneath the north continues to drop. China's Ministry of Water Resources has acknowledged that roughly 80 percent of shallow groundwater in the North China Plain is unfit for drinking or bathing, contaminated by decades of industrial waste and agricultural runoff. The Yellow River, the "Mother River" that fed Chinese civilization for five thousand years, stopped reaching the sea for 226 days in 1997. The river that built an empire ran dry before it hit the ocean.

And the Tibetan glaciers that feed the Yangtze, the Yellow, the Mekong, the Indus, and the Brahmaputra are melting. Short term, more meltwater. Long term, the source dries up, and with it the water supply for roughly a billion people across China, India, and Southeast Asia. China is building railroads across three continents while its own water supply is collapsing underneath it.

The Middle East is where water scarcity is farthest advanced and where one country has proven it is solvable. Iran, Iraq, Jordan, Syria, and the Gulf states are all drawing down aquifers that recharge slowly or not at all. The Tigris and Euphrates are shrinking as Turkey dams the headwaters. Syria's worst recorded drought preceded its civil war. Jordan is one of the most water-stressed countries on Earth.

And then there is Israel, a country in the middle of a desert with no significant natural freshwater resources, which solved the problem through sheer investment. Israel built five major desalination plants along its Mediterranean coast, invested aggressively in drip irrigation and water recycling, and turned a water-scarce nation into a water exporter. Israel now produces more freshwater than it consumes, with the surplus flowing to Jordan and the Palestinian territories. The technology is proven. The engineering is understood. The cost is manageable for any industrial economy.

Desalination is to water what nuclear is to energy: a known solution that works at scale, waiting for the political will to deploy it. Chapter 22 will explain how nuclear power makes desalination economically viable even for countries that cannot afford the energy costs of running the plants on fossil fuels. For now, the point is simple: the water crisis has a solution. One country deployed it. The rest are watching their aquifers drop and hoping it rains.

The American version is happening in California. The Salton Sea was created by accident in 1905 when an irrigation canal broke and Colorado River water flooded a desert basin. For decades it was California's largest lake, ringed by resort towns. Then the water that sustained it was diverted to San Diego and Los Angeles under transfer agreements, and the lake began to die. It is shrinking, concentrating salt and toxins, and the exposed lakebed is poisoning the surrounding communities. Pesticide residue, heavy metals, and selenium dust blow across the Imperial Valley, producing some of the highest childhood asthma rates in the state. The communities are overwhelmingly poor and Latino. The state has known about this for decades and done effectively nothing. Soviet irrigation policy destroyed the Aral Sea. American water politics is destroying the Salton Sea. The mechanism is identical: divert the water for economic purposes, let the lake die, and poison the people who live nearby.

The slow-motion version is under the Great Plains. The Ogallala Aquifer, one of the largest underground freshwater reservoirs in the world, irrigates roughly 30 percent of American cropland. It waters the wheat, corn, and cattle operations that make the United States the world's largest food exporter. It is being drained faster than it recharges. In parts of Kansas and Texas, the water table is dropping one to three feet per year. Wells that used to reach water at 100 feet now have to drill to 300. When the Ogallala runs dry, and in some regions it is already functionally depleted, the most productive farmland in the Western Hemisphere becomes dryland. That connects directly to the food chain in Chapter 14: the fertilizer feeds the crops, but the water grows them. Remove either input and the system fails.

And the Ogallala is not the only water crisis in the American West. Tree ring studies going back 1,200 years show that the megadrought that began around 2000 in the southwestern United States is the worst in at least that period. Lake Mead, the reservoir behind Hoover Dam that supplies water to 25 million people across Nevada, Arizona, and Southern California, dropped to 22 percent of capacity in 2022 before modest recovery. Lake Powell, the other major Colorado River reservoir, was even worse.

The Colorado River Compact, the legal framework that allocates water among seven states and Mexico, was negotiated in 1922 during one of the wettest periods in the region's recorded history. It allocated more water than the river carries in a normal year, let alone a drought year. The math has never worked, and climate change is making it worse. The states are fighting over shares of a river that does not contain enough water to honor the promises that were made a century ago. Phoenix, Las Vegas, Tucson, and much of Southern California's agriculture depend on a water source that is shrinking, over-allocated, and governed by a legal framework built on a lie.

Climate models project that aridification of the American Southwest will continue and intensify through the rest of this century. This is not a drought in the traditional sense, a temporary dry spell that ends when the rains return. This is a permanent shift in the region's hydrology. The rains are not coming back. The 40 million people who live in the Colorado River basin will have to adapt to less water, permanently, while the political system that allocates the water has not updated its assumptions since Warren Harding was president.

And then there is the water coming out of American taps. The EPA estimates 9.2 million lead service lines still deliver drinking water to American homes. Flint, Michigan became the national symbol, but Jackson, Mississippi's water system collapsed in 2022 and still hasn't been fully repaired. PFAS, the "forever chemicals" used in nonstick coatings and firefighting foam, contaminate water systems serving over 100 million Americans. The EPA set enforceable limits in April 2024, decades after the contamination was identified. The infrastructure that delivers clean water to American homes is aging, underfunded, and in many communities already failing. The country that cannot deliver clean water to Flint, cannot fix the Salton Sea, is draining the aquifer that irrigates 30 percent of its cropland, and is watching the river that supplies 40 million westerners disappear, is not a country prepared to manage the water crises of the next fifty years.

• • •

The Arctic is the geographic transformation that most people don't think about, and it may matter more than anything else in this chapter.

As ice retreats from the Arctic Ocean, shipping lanes open. The Northern Sea Route along Russia's coast is already commercially viable for part of the year and becoming more so. A fully navigable Arctic reduces the shipping distance from Asia to Europe by roughly 40 percent compared to the Suez Canal route. That's a revolution in global trade logistics, comparable to the opening of the Suez or Panama Canals.

Whoever controls the Arctic coast controls the new trade route. Russia has the longest Arctic coastline and has been building military bases, icebreaker fleets, and port infrastructure along it for years. China, which has no Arctic coastline, has declared itself a "near-Arctic state" and is investing heavily in Arctic shipping and research. The partnership writes itself: Russia has the coast, China has the money and the ships. The Belt and Road extends north.

The Arctic also holds enormous untapped energy and mineral resources. Oil and gas deposits that were previously inaccessible under ice are becoming reachable. Rare earth minerals, critical for the energy transition, exist in Arctic deposits. The competition for Arctic resources is quiet now but will intensify as the ice retreats and the accessibility increases.

For the United States, the Arctic is both an opportunity and a vulnerability. Alaska provides an Arctic coastline and a claim to Arctic resources. But the US has underinvested in Arctic infrastructure for decades. The icebreaker fleet is a fraction of Russia's. The military presence is minimal compared to what Russia has built. The strategic attention has been focused on the Pacific and the Middle East while Russia and China have been positioning for the Arctic. Another case of looking in the wrong direction while the map changes underneath.

. . .

Here's the variable that most geopolitical analysis completely misses: climate introversion.

Every force in this chapter, the agricultural shifts, the water crises, the sea-level displacement, the extreme weather, requires a response. And the response costs money, attention, and institutional capacity. Rebuilding coastlines. Relocating populations. Restructuring agriculture. Managing internal migration. Redesigning infrastructure for conditions it wasn't built for. Responding to disasters that used to happen once a generation and now happen every few years.

If a major power is spending 5 to 10 percent of its GDP on climate adaptation by mid-century, and some estimates suggest that's where the numbers are heading, then the surplus available for military spending, foreign investment, and power projection shrinks dramatically. You can't build aircraft carriers and sea walls at the same time if you don't have the money for both.

This undermines the bipolar model that Part Seven proposes. The model assumes the US and China both have the resources to dominate their respective hemispheres. But what if both are so consumed by domestic climate management that they can barely hold their own countries together, let alone manage global spheres of influence?

China faces severe water scarcity in the north, devastating flooding in the south, and a coastline where hundreds of millions of people live in areas vulnerable to sea-level rise. Shanghai, Guangzhou, Shenzhen, Tianjin, the economic engines of the Chinese miracle, are all coastal cities at risk.

Protecting them requires infrastructure investment on a scale that competes directly with military spending and foreign expansion.

The United States faces western drought, coastal flooding on both seaboards, increasingly destructive hurricanes, wildfire seasons that have become year-round in parts of the West, and agricultural disruption in regions that feed the world. None of this is existential for a country with America's resources and geographic depth. But it's expensive. And expensive means trade-offs.

The darkest version of the model isn't two orderly hemispheres dividing the world between them. It's fragmentation. Every country turns inward. The global systems, trade, finance, security, that currently connect things atrophy. Not because anyone chose isolationism, but because nobody can afford to maintain the connections when the home front is on fire. A world of fortress states, each managing its own climate catastrophe, with the spaces between becoming ungoverned chaos.

That's the worst case, not the base case. But it's worth naming because the optimistic projections assume that the major powers retain enough capacity to manage both domestic adaptation and external engagement. If climate costs spiral beyond projections, that assumption breaks.

. . .

Climate migration is the force that ties all of this together and connects it to the rest of the book.

When land stops growing food, people move. When coasts flood, people move. When water runs out, people move. When heat becomes lethal, people move. This is already happening at the community scale. In Newtok, Alaska, a Yup'ik village of 400 people has been relocating for over a decade because the

Ninglick River, accelerated by permafrost thaw, has been eating the village out from under them. In 2014, the Fijian village of Vunidogoloa became the first community in the world to complete a government-assisted relocation due to sea-level rise, moving uphill after years of flooding that made their homes uninhabitable. Off Panama's Caribbean coast, the indigenous Guna people of Gardí Sugdub, a coral island barely a meter above sea level, began relocating to the mainland in 2024 after decades of worsening tidal floods. These are not distant projections. They are people packing boxes. The scale of movement that climate change will produce over the next fifty years dwarfs anything in human history. We're not talking about refugee caravans of thousands. We're talking about population relocations of hundreds of millions.

Some of that movement will be internal: rural to urban, south to north within countries, coast to interior. Some will cross borders. The pressure falls heaviest on the temperate countries of the Northern Hemisphere, which are both the most attractive destinations and the societies already struggling with demographic decline. The cruel irony: the countries that need immigration to solve their demographic crisis are the same countries that will face massive climate migration pressure, and the political systems of those countries are treating immigration as a threat rather than a lifeline.

Chapter 32 will explore the migration century in detail. For now, the point is that climate change doesn't just add a crisis on top of the geopolitical competition. It reshapes the physical geography the competition is fought over, redirects the flow of people and resources, and imposes costs that constrain every other strategic choice.

The map is not fixed. It's moving. And the countries that plan for the map of 2050 rather than the map of 2025 will have a decisive advantage over the ones that don't.

China is planning for it. The Belt and Road, the Russian Far East play, the Arctic investment, all of it is positioning for the climate-altered geography of the future.

The United States is fighting a war over a shipping lane it doesn't need, in a region that's becoming less habitable every year, while the actual strategic landscape shifts underneath its feet.

And there's one more horseman still to come. The one that arrives without warning, hits the weakened system at its most vulnerable point, and reveals every fragility that the other three horsemen created.

Chapter 16: The Next Plague

I was involved in pandemic preparation work before COVID. Raised the alarm. Got laughed out of the room. Then COVID hit and it was exactly the mess I expected, for exactly the reasons I had laid out. Enormous amounts of money spent on very little that actually worked. Supply chains exposed. Stockpiles nonexistent. Institutional coordination a disaster. And then the moment it was over, the lessons were filed away and the preparation budgets were cut again. We are not ready for the next one. Not even close. When it comes, and it will come, it is going to be quite a shock to a country that had the chance to learn and chose not to.

In 2018, a virus you have probably never heard of killed roughly 300 million pigs in China. African Swine Fever swept through the Chinese herd, which was the largest on Earth, and destroyed nearly half of it within two years. Pork prices doubled. China, which consumes more pork than any country in the world, had to buy from the global market at emergency rates, distorting protein prices across three continents. The government culled millions of animals, sometimes burying them alive in mass graves. The outbreak was the largest destruction of livestock in recorded history. It was a pandemic. It killed no humans. And it shattered a food system that fed 1.4 billion people.

Most Americans never noticed. They should have, because the lesson of African Swine Fever is the lesson this chapter is built around: pandemics don't just kill people. They kill food systems. And the food systems are already under the stress that Chapter 14 described.

African Swine Fever has not been eradicated. It is still circulating in Asia, Europe, and Africa. It reached the Dominican Republic in 2021, the closest it has come to the US mainland. There is no vaccine. There is no treatment. The only response is mass culling. And ASF is not alone.

• • •

H5N1 avian influenza has been spreading through US poultry flocks for years, killing tens of millions of birds and driving egg prices to record levels. In early 2024, it jumped to dairy cattle, a transmission route that virologists had not anticipated. By 2025, H5N1 had been confirmed in dairy herds across multiple states. In March 2024, a farmworker in Texas tested positive for H5N1 after handling infected cattle, the first confirmed bovine-to-human transmission in the United States. He recovered.

The case fatality rate for H5N1 in humans, across all recorded cases worldwide, is roughly 50 percent. Not 1 percent, like COVID. Fifty. The only thing standing between H5N1 and a civilizational catastrophe is that the virus has not yet acquired the mutations needed for efficient human-to-human transmission.

Virologists estimate that a small number of changes in the hemagglutinin protein, perhaps as few as three to five mutations, could enable airborne spread between people. Those mutations could occur naturally in any mammalian host where avian and human influenza strains co-infect. Dairy cows, with millions of people in close contact with them daily, are a plausible mixing vessel. This is not alarmist speculation. It is the consensus of the virological community, expressed in publications going back twenty years. The question is not whether H5N1 can become a human pandemic. The question is whether it will, and when.

The global food system is under constant biological attack from pathogens that most people have never heard of. Wheat blast, a fungal disease caused by Magnaporthe oryzae, has devastated wheat harvests in Bangladesh and is spreading to new regions as climate change expands the fungus's range. Tropical Race 4, a soil-borne fungal pathogen, is killing the Cavendish banana, which accounts for 99 percent of global banana exports. There is no fungicide that stops it. There is no resistant commercial variety available at scale. The banana your children eat may not exist in twenty years. Citrus greening, spread by a tiny insect called the Asian citrus psyllid, has destroyed Florida's orange industry, cutting production by more than 80 percent since 2005, and is spreading to every citrus-growing region on Earth. These are not hypothetical threats. They are active pandemics. They are happening now. The difference between them and COVID is that they kill crops and livestock instead of people, and the media doesn't cover them because dying pigs and rotting bananas don't generate the same engagement as overflowing hospitals.

Biosecurity is food security. That sentence should be written on the wall of every agriculture department and every national security agency in the world, and it is written on none of them. The food chain from Chapter 14 doesn't just face depletion from resource scarcity. It faces destruction from biological threats that are evolving, spreading, and accelerating faster than the response systems can track them.

And there is a slower pandemic that has been building for decades and that nobody calls a pandemic because it kills too quietly. Antimicrobial resistance, the evolution of bacteria that no longer respond to antibiotics, killed an estimated 1.27 million people in 2019 and is projected to kill 10 million annually by 2050 if current trends continue. More than cancer. The mechanism is simple: every time antibiotics are used, the bacteria that survive are the ones that resist the drug. Those resistant bacteria reproduce and spread. The more antibiotics are used, the faster resistance develops.

And antibiotics are massively overused, in human medicine where doctors prescribe them for viral infections they cannot treat, and in agriculture where livestock are dosed with antibiotics not to cure disease but to promote growth in the confined, filthy conditions of factory farming. The agricultural use alone accounts for roughly 70 percent of all antibiotics consumed in the United States. The antibiotics that are supposed to save your life during surgery or chemotherapy are being fed to chickens to make them grow faster, and the resistant bacteria that emerge from those chickens enter the food supply, the water supply, and the soil.

When a post-surgical infection no longer responds to any available antibiotic, the patient dies of something that has been treatable since 1943. We are breeding our way back to the pre-antibiotic era, and the pace is accelerating.

• • •

COVID-19 killed roughly 7 million people by official count, probably closer to 20 million by excess mortality estimates. It was not the big one. It was the dress rehearsal, and we failed it. The infrastructure that was supposed to protect against the next human pandemic is in worse shape now than when COVID hit, because we did not learn from the rehearsal. We watched it happen. We lived through it. And we changed nothing.

Zoonotic spillover, the process by which a pathogen jumps from an animal to a human, is increasing for structural reasons that are all getting stronger. Climate change pushes animal populations into new ranges, bringing reservoir species into contact with humans they have never encountered. Deforestation in the tropics destroys the habitat buffer between wildlife and settlements. Urbanization concentrates people in dense megacities that function as incubators for respiratory disease. A novel pathogen that emerges in rural Southeast Asia reaches every continent within 48 hours through commercial aviation. And factory farming, the system that produces most of the world's meat, packs thousands of genetically similar animals into confined spaces dosed with antibiotics that breed resistant bacteria, creating conditions where a novel influenza strain could emerge and spread before anyone detects it.

. . .

The second reason is scarier than the first, because it involves human agency rather than nature.

Bioengineering capability is democratizing at a pace that should terrify anyone paying attention. CRISPR gene editing technology, which allows precise modification of DNA, costs a few hundred dollars and can be ordered online. The knowledge to modify pathogens, to make a virus more transmissible, more virulent, or more resistant to treatment, is increasingly accessible through published scientific literature and AI-powered research tools.

This is the dual-use problem at its most acute. The same AI that can model protein structures and predict how a virus will mutate, which is fantastic for designing vaccines and treatments, is equally capable of modeling how to engineer a more dangerous pathogen. The same CRISPR tools that let a lab develop a gene therapy for cancer let a different lab modify a flu virus to evade existing immunity. The line between defensive research and offensive capability is thin enough that multiple countries are probably straddling it right now.

Gain-of-function research, the practice of deliberately making pathogens more dangerous to study how they might evolve naturally, is conducted in dozens of laboratories worldwide with varying levels of biosafety. The debate about whether COVID originated from a natural spillover or a laboratory incident remains unresolved, but the question being plausible at all tells you something about the state of biosafety in research facilities handling dangerous pathogens.

Whether COVID came from a lab or a market, the next pandemic could come from either.

And then there's the deliberate scenario. A state actor that has vaccinated its own population, or developed effective treatment protocols, could release a pathogen as a strategic weapon. Or it could exploit a natural pandemic by hoarding vaccines, withholding treatments, or using pandemic chaos as cover for military action. Attribution of a bioweapon attack is nearly impossible because the evidence looks identical to a natural outbreak. The plausible deniability is built into the weapon itself.

• • •

The response infrastructure, the system that is supposed to protect us when a pandemic hits, is in worse shape than it was before COVID. National stockpiles of antivirals and PPE were drawn down and not fully replenished. This pattern is the same as the missile stockpile: crisis depletes the inventory, post-crisis budget pressures prevent restocking, the next crisis finds the cupboard bare. COVID turned public health into a partisan battlefield, and the institutional trust that pandemic response depends on was destroyed. Half the American population would refuse to comply with public health measures during the next pandemic regardless of the mortality rate. That is not a guess. It is the observed behavior during COVID applied to a future scenario.

A pathogen that kills 5 percent of infected people would hit a society where 40 percent of the population refuses to take precautions, where the healthcare system has been hollowed out by private equity, where the supply chains for critical drugs run through China, and where the political leadership is more likely to argue about the pathogen's origins on social media than to coordinate a response. The dress rehearsal showed all of these weaknesses. None of them have been fixed.

The pharmaceutical supply chain vulnerability deserves its own paragraph because it connects the pandemic threat to the broader theme of strategic dependency that runs through this book. Over 80 percent of the active pharmaceutical ingredients used in American drugs are manufactured overseas, primarily in China and India. Generic antibiotics, blood pressure medications, pain relievers, and the precursor chemicals for dozens of critical treatments come from Chinese factories that operate under a regulatory framework the FDA cannot inspect with any consistency. During COVID, India restricted exports of certain drugs to preserve domestic supply. China controls the global supply of heparin, a blood thinner used in every hospital in America, produced from pig intestines processed in Chinese facilities. A pandemic that disrupts Chinese manufacturing, or a geopolitical crisis that prompts China to restrict pharmaceutical exports, would leave American hospitals unable to treat patients regardless of how many ventilators or ICU beds they have. The drugs wouldn't arrive. This is the Hormuz problem applied to healthcare: the supply chain runs through a chokepoint controlled by a potential adversary, and nobody has built an alternative.

• • •

Now layer the pandemic on top of the other three horsemen and the geopolitical situation.

A pandemic during an active multi-theater conflict is catastrophically destabilizing. The USS Theodore Roosevelt had to pull into port during COVID because the virus was tearing through the crew. Imagine that with a pathogen that kills at fifty times the rate, during a shooting war where pulling into port is not an option.

A pandemic accelerates the demographic crisis. A severe pandemic that kills 5 to 10 percent of the over-65 population in countries already short on workers is an economic catastrophe hitting the weakest point of the weakest economies.

A pandemic shatters the food supply chain that Chapter 14 described. Workers can't reach processing plants. Shipping stops. Fertilizer deliveries stall. If that happens during planting season, the result is not a temporary shortage. It is famine.

A pandemic forces the same introversion that Chapter 15 identified for climate. Borders close. Trade stops. International cooperation collapses. Every country focuses on its own survival. The global systems that connect economies atrophy in weeks rather than the decades that climate introversion takes. A severe pandemic could produce the fortress-state fragmentation scenario faster than any other force in this book.

●　●　●

The pharmaceutical dimension adds a strategic layer. If Beijing can decide who gets antibiotics and who doesn't, who gets vaccine precursors and who waits, that is leverage that makes the Strait of Hormuz look like a parking dispute. Every country will restrict pharmaceutical exports during the next pandemic, the same way India restricted them during COVID. The global pharmaceutical supply chain, like the fertilizer supply chain and the energy supply chain, is optimized for efficiency in normal times and collapses under stress. The redundancy that would make it resilient was stripped out because redundancy is expensive and the market rewards efficiency.

Sound familiar? The *Enshittification* pattern again. Optimize for short-term efficiency. Strip out redundancy. Extract maximum value from the lean system. Then watch the lean system shatter when it encounters a stress that the optimization didn't account for. The defense industrial base, the healthcare system, the food supply chain, the pharmaceutical supply chain, all optimized to the same standard and all vulnerable to the same kind of shock.

• • •

The optimistic case, because this book always makes the optimistic case, is that the tools to survive the next pandemic are better than what we had for COVID. Dramatically better.

mRNA vaccine technology, which was developed for COVID in record time, is now a platform that can be adapted to new pathogens in weeks rather than years. The fundamental breakthrough, using messenger RNA to instruct cells to produce antigens that trigger immune response, works for any pathogen whose protein structure can be decoded. AI can decode protein structures faster than any human research team. The combination of mRNA platforms and AI-powered pathogen analysis means that vaccine development for the next pandemic could begin within days of the pathogen being identified.

AI surveillance systems can monitor disease outbreaks in real time, tracking case reports, wastewater data, hospital admissions, and genomic sequences to identify emerging threats before they become pandemics. The early warning capability exists. Whether the political systems will act on early warnings, given that they ignored the warnings for COVID, is the question that technology can't answer.

And here's where this chapter connects to the later chapters on human enhancement. Enhanced humans with integrated biosensors, AI-linked health monitoring, and direct neural access to medical data would survive a pandemic far better than baseline humans. They'd detect infection earlier, respond to treatment faster, and coordinate containment more effectively. A pandemic that kills 10 percent of baseline humans might kill a fraction of a percent of enhanced humans. That's a survival advantage that would accelerate the enhancement adoption curve faster than any marketing campaign. Nothing motivates people to embrace new technology like the fear of death.

But that advantage creates its own problem, which Chapter 25 will explore: a world where enhanced populations in wealthy countries survive plagues that devastate unenhanced populations everywhere else. The pandemic becomes another mechanism of divergence, another force pulling the two tiers of humanity further apart.

· · ·

Part Four described four horsemen: demographic collapse, resource depletion through the oil-fertilizer-food chain, climate change, and pandemic risk. None of them is a standalone crisis. Each one interacts with the others to create compound effects that are worse than the sum of their parts. Demographic decline produces aging societies that are more vulnerable to pandemics. Climate change drives zoonotic spillover that increases pandemic risk. Resource depletion drives food insecurity that makes populations less resilient to every kind of shock. Pandemics disrupt the food chains and economic systems that were already under stress from the other three.

It's a system. And the system is under pressure from all directions at the same time.

The temptation, at this point, is despair. Four horsemen. Interconnected. Self-reinforcing. Getting worse. But despair is a luxury we can't afford, because despair is paralyzing and the situation requires action.

Before we get to the tools, though, there's one more piece of the puzzle. The force that's been quietly building behind all of these crises, positioning itself to benefit from every one of them, while the country that should be preparing for the future exhausts itself in wars it doesn't need to fight.

The quiet empire. The one that doesn't fire shots. The one that builds railroads.

If Part Four left you wondering whether any of these problems are solvable, hold that thought. Every one of them is. A single technology solves the energy crisis, the food crisis, and the water crisis at the same time. Another rewrites the demographic model. A third makes pandemics survivable. The tools are coming in Part Six. But first you need to understand who else is watching these crises unfold, and what they are building while the United States stares in the wrong direction.

Welcome to Part Five.

Part Five: The Quiet Empire

Chapter 17: The Matador's Cape

In a bullfight, the matador uses a large red and magenta cape called a capote during the opening passes. He swings it wide, drawing the bull's charge to the cape while his body stands to the side. The bull sees the cape, not the man. It commits everything to the cape, horns lowered, full speed, absolute conviction that the target is where the fabric is moving. The matador steps aside. The bull hits nothing. And it keeps charging the cape, over and over, because the cape is all it can see.

For twenty years, the United States has been the bull. Taiwan is the cape.

. . .

Every Pentagon war game about China centers on Taiwan. The most extensive unclassified simulation, run by the Center for Strategic and International Studies in 2023, played the scenario 24 times with different assumptions. The conclusion: a Chinese amphibious invasion would probably fail. The cost of that failure, for everyone, would be catastrophic. The United States loses two aircraft carriers, 10 to 20 major surface warships, 200 to 400 aircraft, and roughly 3,200 troops in three weeks of fighting. China loses 138 warships, 155 combat aircraft, and roughly 10,000 soldiers killed in action or drowned. Taiwan's entire navy of 26 destroyers and frigates goes to the bottom. The island's economy is shattered, its electricity grid destroyed, its military degraded beyond near-term recovery.

Mark Cancian, the CSIS senior adviser who led the project, put the result bluntly: the United States wins the battle but suffers more in the long run than the defeated China. A Pyrrhic victory that damages American global military positioning for a decade.

Every congressional hearing about Pacific strategy focuses on Taiwan. Every National Defense Strategy identifies the defense of Taiwan as the pacing scenario for force structure decisions. The ships are designed around it. The planes are bought for it. The Marines were reorganized for it. The bases in Japan and Guam are positioned for it. Billions of dollars in annual defense spending are justified by it.

But a full-scale Taiwan invasion is the least likely scenario, not the most likely. The range of plausible confrontations between "nothing happens" and "D-Day across the Taiwan Strait" is where the actual risk lies, and the Pentagon's war games barely touch it. A Chinese coast guard blockade of a Philippine-occupied shoal in the South China Sea, squeezing resupply until Manila blinks or calls for American help. A confrontation over the Senkaku Islands that forces Japan to invoke Article 5 of the US-Japan security treaty for the first time. A "quarantine" of Taiwan that stops short of invasion but interdicts shipping, cuts undersea cables, and dares the US to fire the first shot. A cyberattack that disables Taiwan's power grid during an election crisis and creates the conditions for a political capitulation that never requires a single PLA soldier to get wet.

Each of these scenarios is more likely than an amphibious invasion in any given year, and each puts the United States in a position where the response options are bad and worse. The force the US has built is optimized for the big war. The confrontations that are coming are the small ones that don't justify a carrier strike group but that shift the balance of power incrementally, one shoal, one cable, one grey-zone provocation at a time.

The question is never whether the US should prepare for a Taiwan contingency. Of course it should. The question is whether the fixation on Taiwan is causing the United States to miss what China is doing.

Because what China is doing is not preparing to invade Taiwan. What China is doing is building the largest empire in human history without firing a single shot. And the US doesn't have a doctrinal response, a budget line, or an institutional framework to counter it, because the entire strategic apparatus is pointed at the cape.

• • •

Let's start with why an invasion of Taiwan is irrational, and why China almost certainly knows it.

Taiwan sits roughly 100 miles off the Chinese coast, across the Taiwan Strait. An amphibious invasion would require transporting hundreds of thousands of troops, plus their equipment, supplies, and support systems, across 100 miles of open water against a defender that has spent decades preparing for exactly this scenario. Taiwan's beaches are limited and well-fortified. Its terrain inland is mountainous. Its military is specifically designed for anti-invasion defense: mines, anti-ship missiles, mobile coastal defense batteries, hardened shelters, and a reserve force that would fight from urban terrain that an invader would have to clear building by building.

The Taiwan Strait is not the English Channel. D-Day was launched across 20 miles of water against defenders who didn't know exactly where or when the attack was coming. A Taiwan invasion would cross 100 miles against a defender that has been watching the buildup for months, because you cannot assemble an invasion force of that size without it being visible from space. There would be no tactical surprise. China would be committing hundreds of thousands of troops to a contested crossing against prepared defenses, in an age when anti-ship missiles can sink a transport ship from a hundred miles away.

Even if China succeeded, which would require absorbing catastrophic losses, what would it have? An island of 24 million hostile people. A semiconductor industry that would not survive the battle. TSMC, the world's most advanced chipmaker and the real prize that people think China wants, has planned for the invasion scenario. Chairman Mark Liu told CNN flatly: "Nobody can control TSMC by force. If there is a military invasion you will render TSMC factory non-operable." ASML, the Dutch company that makes the irreplaceable extreme ultraviolet lithography machines TSMC needs to produce cutting-edge chips, has reportedly equipped them with remote kill switches that can disable them from the Netherlands. The most advanced fab equipment is irreplaceable in the short term and incredibly fragile. Bombing doesn't destroy it. An EMP doesn't destroy it. But cutting the power, severing the ultrapure water supply, or evacuating the technicians who know how to operate it renders it useless. The factory might survive an invasion intact. The capability to run it would not.

So China would spend a trillion dollars, absorb a million casualties, endure global sanctions that would make the Russian experience look mild, trigger a possible military confrontation with the United States, and at the end of it all possess a smoking ruin populated by people who hate them, with a semiconductor industry that no longer functions. The cost-benefit analysis doesn't work. Not even close.

• • •

180

Chinese strategists know this. They're not stupid. The PLA's own internal assessments, to the extent that outside analysts can reconstruct them, suggest a more careful approach than the Normandy-style invasion that Western war games obsess over. A blockade rather than an invasion. A quarantine rather than an assault. Political coercion backed by military threat rather than the military operation itself. Or, most likely, continued patience: let the demographic and economic trends play out, let Taiwan's own internal politics evolve, let the cross-strait economic integration that already exists deepen, and wait for a moment when unification can be achieved through negotiation, coercion, or fait accompli rather than force.

China has been waiting for Taiwan since 1949. The Chinese Communist Party thinks in decades and centuries, not in election cycles. An invasion that risks everything, destroying the prize in the process, when patience offers a less costly path, is not the strategic culture's preferred approach. Sun Tzu wrote that the supreme art of war is to subdue the enemy without fighting. That's not just a famous quotation. It's operational doctrine.

But here's what matters for this book: whether or not China invades Taiwan, the fixation on Taiwan serves China's interests perfectly.

• • •

Every dollar the United States spends preparing for a Taiwan invasion scenario is a dollar not spent countering Chinese economic expansion across Eurasia. Every war game that focuses on the First Island Chain is a war game that ignores the Belt and Road. Every Marine battalion reorganized for Pacific island-hopping is a battalion not available for anything else. Every congressional hearing about Taiwan is a hearing that doesn't happen about Chinese infrastructure investment in Central Asia, or Chinese acquisition of rare earth processing capacity, or Chinese penetration of African economies, or Chinese positioning in the Russian Far East.

Taiwan is the matador's cape. China waves it because the bull keeps charging. And while the bull charges the cape, the matador is doing something else entirely.

The demographic clock adds urgency that makes this reading even more compelling. China has maybe 15 to 20 years before the workforce shrinks past the point where sustained imperial expansion is feasible. The window is closing. In that window, the priority is locking in structural advantages that will persist after the demographic decline: resource access, infrastructure dependencies, financial systems, energy transition dominance, and territorial positioning.

Taiwan doesn't serve that priority. Taiwan is a semiconductor island with no natural resources and a hostile population. It would cost a fortune to take and a fortune to hold. It would trigger sanctions that would cut China off from the very global markets it's trying to dominate. And it would consume military resources that are needed for the real priority: securing Eurasia.

Eurasia serves the priority perfectly. Resource access through Belt and Road. Energy security through Russian gas pipelines that can't be interdicted by the US Navy. Agricultural potential in warming Siberian territory. Mineral wealth in Central Asia. Labor access in Africa. Port access around the Indian Ocean rim. All of it acquired through investment, infrastructure, and economic dependency rather than military force. All of it invisible to a strategic establishment that's staring at satellite imagery of Chinese amphibious ships and counting landing craft.

• • •

The Iran war is the most vivid demonstration of how the cape works.

While the United States was pouring military assets into the Middle East, burning through missile stockpiles, wearing out its carrier fleet, and degrading its readiness for the scenario it claims is its top priority, China was doing the following:

Providing Iran with BeiDou satellite navigation to make Iranian missiles harder for the US to jam. Selling radar technology and electronic warfare expertise that enhanced Iran's ability to fight. Sending an intelligence vessel, the Liaowang-1, to the Arabian Sea to observe US naval operations in real time. Stockpiling oil imports at 16 percent above normal rates in the months before the war, indicating advance knowledge or at least advance preparation. Negotiating with Iran for safe passage of Chinese ships through the Strait of Hormuz while everyone else's shipping was stopped. Accelerating the Power of Siberia 2 pipeline negotiations with Russia, deepening the energy relationship that makes China less dependent on maritime supply chains the US could interdict.

And taking notes. On everything.

How fast does the US consume THAAD interceptors? How quickly can the industrial base ramp production? What happens to carrier operations after 300 days at sea? How do the Aegis systems perform against saturation attacks? Where are the electronic warfare gaps? How do the logistics work when the Strait of Hormuz is closed? What do the allies do when they get hit, do they stay in or pull out? How does American public opinion respond to casualties?

Every one of those data points is gold for a military planning to fight the United States in the Pacific. And China got all of it for free, without risking a single soldier, because the US fought the war for them while they watched.

• • •

The US strategic establishment is not entirely blind to this. There are analysts who write about the Belt and Road as a strategic threat. There are think tanks that study Chinese economic penetration of Central Asia and Africa. There are intelligence officers who track Chinese technology transfer to adversaries like Iran. The information exists.

What doesn't exist is a response framework. The American military is built to counter kinetic threats: invasions, missile launches, naval confrontations. It has tools for those things. Carrier strike groups. Missile defense batteries. Stealth bombers. Special operations forces. All of them designed to find, target, and destroy things that are trying to kill you.

There is no weapon system that counters a Chinese-built railroad in Kazakhstan. There is no military response to a Belt and Road loan that gives Beijing control of a port in Sri Lanka. There is no Pentagon program designed to compete with Chinese investment in African telecommunications infrastructure. The threats are economic, not kinetic. The tools don't exist because the institution that develops tools is a military, and militaries develop military tools.

The State Department, which should be the institution countering economic statecraft, has been starved of resources for decades. USAID, which should be offering development alternatives to Belt and Road, has a budget that is a rounding error compared to Chinese investment. The economic tools that could compete, export financing, infrastructure investment, trade agreements, technology partnerships, are scattered across multiple agencies with no unified strategy and no senior leadership that views them as strategic weapons.

China has a unified strategy. It's called the Belt and Road. It has senior leadership. It's called Xi Jinping. It has funding. It's measured in trillions. And it has institutional coherence: the entire Chinese state apparatus, from the military to the banks to the construction companies to the diplomatic corps, is aligned toward the same set of objectives.

America has the Pentagon staring at Taiwan, the State Department underfunded and demoralized, USAID under threat of elimination, the intelligence community spread across sixteen agencies that don't talk to each other, and a political system that can't agree on a budget for more than a few months at a time.

The bull charges the cape. The matador builds an empire.

• • •

The counter-strategy is not military. You cannot deter a railroad with an aircraft carrier. You cannot sanction a port that a country built with its own money on land it leased through a bilateral agreement. The tools that would counter China's empire-building are economic: competing infrastructure investment, trade agreements that offer alternatives to Chinese dependency, development financing that doesn't come with Chinese strings, digital infrastructure that doesn't route through Chinese equipment. The US has none of these at scale. It tried with the Build Back Better World initiative under Biden, which was renamed the Partnership for Global Infrastructure and Investment, which produced announcements and photo opportunities and almost no actual infrastructure. The US government is institutionally incapable of competing with China in this domain because the competition requires patient, sustained investment over decades, and American political cycles reward short-term spending and punish long-term commitment. China's advantage is not money. It is time horizon. And a democracy that changes direction every four years cannot compete with an autocracy that plans in decades.

Together, they describe a power that has studied the board more carefully than its opponent, positioned its pieces more patiently, and played the game at a level the opponent doesn't even recognize as a game.

The bull is still charging the cape. But the fight is almost over. And the bull doesn't know it yet.

Chapter 18: Railroads, Not Rifles

In the 19th century, the British Empire controlled a quarter of the world's land surface. It did this not primarily through military conquest, though there was plenty of that, but through a system of commercial dominance backed by infrastructure. The British built railroads in India, ports in East Africa, telegraph lines across oceans, and banking systems that denominated global trade in pounds sterling. The military existed to protect the commercial system, not the other way around. When the costs of maintaining the military exceeded the returns from the commercial system, the empire collapsed. The infrastructure stayed.

China is building the 21st-century version of the British Empire. Except it's doing it faster, across a larger geographic area, and without the military costs that eventually bankrupted London.

. . .

The Belt and Road Initiative, launched by Xi Jinping in 2013, is the largest infrastructure investment program in human history. Numbers are genuinely difficult to comprehend. Over a trillion dollars committed across more than 140 countries. Railroads, ports, highways, power plants, telecommunications networks, fiber optic cables, industrial parks, and special economic zones spanning Asia, Africa, Europe, the Middle East, and Latin America.

But the trillion-dollar headline hides a program under strain. New BRI lending has collapsed since 2020, falling by roughly two-thirds from its peak as recipient countries defaulted on loans and Beijing pulled back. Sri Lanka's Hambantota port became the symbol of the debt trap, but it is not unique. Zambia defaulted on its Chinese debt in 2020. Pakistan has been restructuring BRI-related obligations for years. Ecuador, Laos, and several African nations are in various stages of distress on Chinese loans.

Beijing has been forced to write off bad debts, reschedule payments, and accept that many of the investments will never return what was promised. The empire is still being built. But the construction has slowed, the costs are mounting, and the model that was supposed to demonstrate Chinese superiority over Western aid is producing its own catalog of failures.

This does not mean the BRI has failed. It means the BRI is expensive, imperfect, and running into the same limits that every imperial infrastructure project in history has encountered. The difference between the BRI and, say, British colonial railroads in India is that China is hitting those limits faster because it is building faster and borrowing more aggressively to do it.

The Western media narrative about Belt and Road usually focuses on debt traps: China lends money for infrastructure that the borrowing country can't afford, then seizes the asset when the country defaults. The Hambantota port in Sri Lanka is the standard example. China lent Sri Lanka over $1.4 billion to build the port, which opened in 2010 and attracted almost no commercial traffic. By 2017, Sri Lanka couldn't service its debts. The solution: a 99-year lease of the port to China Merchants Port Holdings for $1.12 billion, which went straight to debt repayment. China got a deep-water port on the Indian Ocean, 200 nautical miles from the sea lanes that carry 80 percent of China's oil imports. Sri Lanka got a receipt. And the debt trap narrative isn't wrong. It's just incomplete. It misses the point the same way describing a chess game as "moving wooden pieces on a board" misses the point.

Belt and Road is strategic colonization dressed in economic language. The infrastructure creates dependency. Dependency creates influence. Influence becomes control. Not the colonial control of the 19th century, with governors and garrisons and flags. Something subtler and more durable: the control that comes from being indispensable.

When a country's railroad runs on Chinese-built track, maintained with Chinese parts, operated with Chinese-trained engineers, financed by Chinese loans, the country doesn't need to be told what to do. It already knows. You don't bite the hand that keeps your economy connected to the outside world. You vote the way Beijing wants at the UN. You give Chinese companies preferential treatment on contracts. You look the other way on human rights. You quietly align your foreign policy with Chinese interests, not because you were conquered, but because the alternative is economic isolation.

• • •

Central Asia is the clearest example of the process in action, and it connects directly to the Russian decline described in Chapter 19.

The five Central Asian republics, Kazakhstan, Uzbekistan, Turkmenistan, Tajikistan, and Kyrgyzstan, were Soviet satellites until 1991. After independence they remained in Russia's orbit: dependent on Russian markets, Russian energy transit infrastructure, and Russian security guarantees. Moscow treated them as a backyard, extracting resources and maintaining influence through a combination of economic ties and the implicit threat that Russia was the only major power that cared enough about the region to intervene.

China moved in with money. Not soldiers. Not threats. Money and concrete and steel.

The China-Central Asia gas pipeline, completed in stages from 2009 to 2014, runs three parallel lines from Turkmenistan through Uzbekistan and Kazakhstan to China's

Xinjiang region, with a combined capacity of 55 billion cubic meters per year---roughly equivalent to the old Nord Stream 1 that carried Russian gas to Europe. Built by China National Petroleum Corporation at a cost exceeding $17 billion in loans and equity, it broke Russia's monopoly on Central Asian energy transit overnight. Before 2009, nearly 70 percent of Turkmenistan's gas exports transited through Russian pipelines. Suddenly Turkmenistan had an alternative customer for its gas, one that paid more and demanded less. Kazakhstan got a transit fee. Uzbekistan got infrastructure investment. All three countries gained economic options they didn't have before, and every one of those options pulled them closer to Beijing and further from Moscow.

Chinese-built railroads are extending across the region. The China-Kyrgyzstan-Uzbekistan railroad, under construction, will create a direct rail link from Chinese manufacturing centers to Central Asian and eventually European markets. Chinese telecommunications companies, primarily Huawei, have built the cellular and internet infrastructure across the region. When your phone call routes through Chinese-built equipment and your internet traffic flows through Chinese-laid fiber, the surveillance implications are significant, but the dependency implications are more significant still.

The Digital Silk Road is the dimension of Belt and Road that Western analysts were slowest to recognize and that may ultimately matter more than the railroads and ports. Huawei has built 4G and 5G networks in over 170 countries. Chinese companies have laid or are building undersea fiber optic cables that carry international internet traffic. Chinese firms supply the cloud computing infrastructure, the data centers, the smart city platforms, and the e-government systems that are being adopted across the developing world.

When a country adopts Chinese telecommunications infrastructure, it is not just buying equipment. It is integrating its communications, its data flows, its government systems, and in many cases its surveillance capabilities into a technology ecosystem that Beijing can monitor, influence, and if necessary disrupt. The United States spent decades building the global internet on American infrastructure and American standards, and then watched as Huawei offered the same services at half the price with Chinese government financing and no questions about human rights. By the time Washington started treating Chinese telecommunications as a security threat and pressuring allies to ban Huawei from their 5G networks, the company had already built the backbone of digital communications across most of Africa, Southeast Asia, Central Asia, and large parts of Latin America.

The ban worked in wealthy democracies that could afford the higher cost of European alternatives. It did not work in the developing world, which is where the competition for the 21st century will be decided.

Russia watched this happen and couldn't stop it. Moscow doesn't have the money to compete with Chinese investment. It doesn't have the construction capacity to build competing infrastructure. Its own economy is being consumed by the Ukraine war. The security guarantee that Russia once provided is looking less credible every month as the Russian military gets chewed up in Eastern Europe. The Central Asian leaders are making the rational calculation: Beijing is the rising power, Moscow is the declining one, and the smart move is to align with the future rather than the past.

. . .

Africa follows the same model at a larger scale, and Chapter 13 already described the geographic constraints that shape what Chinese investment can and can't achieve there. But the investment itself deserves a closer look, because it reveals the strategic logic more clearly than any other region.

China is Africa's largest trading partner and its largest bilateral lender. Chinese state-owned enterprises and private companies have built railroads in Kenya and Ethiopia, ports in Djibouti and Tanzania, highways in Congo and Cameroon, power plants in Nigeria and Angola, and telecommunications networks across the continent. China's first overseas military base is in Djibouti, at the mouth of the Red Sea, positioned to monitor one of the world's most important shipping lanes.

The Western critique of Chinese investment in Africa usually focuses on labor practices and environmental standards, both of which are legitimate concerns, or on neocolonialism, which is ironic coming from the countries that colonized Africa. What the critique misses is the strategic function.

China needs resources. Africa has them. China needs markets. Africa's population is growing while every other continent's is plateauing or declining. China needs labor as its own workforce shrinks. Africa has the youngest population on Earth. The investments aren't charitable. They're not even purely commercial. They're the advance work of an empire that's building the physical infrastructure of dependency across a continent that will matter more in 2050 than it does today.

Russia is playing the same continent with a different playbook. Where China builds railroads, Russia runs information operations. In Africa's Sahel region, Russian-backed disinformation campaigns have weaponized anti-colonial sentiment, using local social media networks to demonize French and Western influence across Mali, Burkina Faso, and Niger. The campaigns worked. France withdrew its military forces from all three countries between 2022 and 2024. Russian paramilitaries, formerly Wagner Group and now reorganized as the Africa Corps, stepped into the vacuum, securing lucrative mining contracts and propping up military juntas friendly to Moscow. The playbook is elegant: use information warfare to expel the competitor, then move in with security forces that double as resource extraction operations. No railroads required. No billion-dollar loans. No decades-long construction projects. Just targeted messaging that exploits genuine grievances, governments that fall into Russia's orbit, and mining rights that flow to Russian-connected firms. China builds dependency through infrastructure. Russia builds it through narrative control. Both models work. Both are expanding across Africa. And neither faces a credible Western counter-strategy.

Whether that dependency translates into the kind of African economic development that the continent needs, as Chapter 13 argued, is doubtful. Chinese investment follows the extraction model: mine to port, resource to ship. It connects Africa to China, not Africa to itself. But from Beijing's perspective, that's the point. China doesn't need an integrated African economy. It needs African resources flowing to Chinese factories, African markets buying Chinese products, and African governments voting with Beijing at the United Nations. The infrastructure is designed to produce exactly those outcomes.

• • •

The energy transition dimension is the one that transforms Belt and Road from a large infrastructure program into a strategic stranglehold.

China produces roughly 80 percent of the world's solar panels. It dominates global battery manufacturing. It controls the processing of the rare earth elements that go into everything from electric vehicle motors to wind turbine generators to missile guidance systems. It has locked up lithium supply contracts in Chile, Argentina, Australia, and the Democratic Republic of Congo. It processes more cobalt than any other country, most of it mined in Congo under conditions that range from exploitative to monstrous.

The new Hormuz isn't a strait. It's a Chinese-owned processing plant in Inner Mongolia.

The West is in the process of trading one energy dependency for another. The old dependency ran through the Persian Gulf and could be disrupted by an Iranian minefield or a Houthi drone. The new dependency runs through Chinese supply chains and can be disrupted by a policy decision in Beijing. What differs is that the old dependency at least involved a resource, oil, that the US could produce domestically. The new dependency involves processing capacity that the US has not built and is not building fast enough to matter.

Every electric vehicle battery produced outside of China still depends on Chinese-processed materials. Every solar panel installation in America and Europe enriches Chinese manufacturers. Every wind farm uses turbines with magnets made from Chinese rare earths. The energy transition that's supposed to free the world from fossil fuel dependency is building a new dependency on Chinese industrial capacity that is in many ways harder to break than the old one.

The US Inflation Reduction Act and similar European programs are trying to build domestic manufacturing capacity for clean energy components. These efforts are real and significant. But they're a decade behind China and they face the same industrial base problems described throughout: workforce shortages, supply chain vulnerabilities, regulatory barriers, and the institutional inertia of a system that was optimized for consumption rather than production.

• • •

The financial architecture is the final layer, and it's the one that will matter most in the long run.

China has been building an alternative to the dollar-based financial system for years. The yuan isn't going to replace the dollar as the world's reserve currency anytime soon; it's not freely convertible and China's capital controls prevent the kind of deep, liquid markets that a reserve currency requires. But China doesn't need to replace the dollar. It just needs to create a parallel system that enough countries use for enough transactions that American financial sanctions lose their teeth.

The Cross-Border Interbank Payment System, CIPS, is China's alternative to SWIFT, the Western-controlled messaging system that enables international bank transfers. CIPS has been growing steadily, adding participants and transaction volume. Belt and Road loans are increasingly denominated in yuan. Chinese digital payment systems, Alipay and WeChat Pay, are spreading across Southeast Asia and Africa. The digital yuan, China's central bank digital currency, is being piloted in cross-border transactions.

None of these individually threatens dollar dominance. Together, over time, they create an alternative ecosystem that countries can use to conduct trade and finance without touching the American banking system. For countries under US sanctions, or afraid of being sanctioned, that alternative is increasingly attractive. Iran trades with China outside the dollar system. Russia, forced out of SWIFT, conducts energy transactions with China in yuan. Every country that joins the alternative system weakens the dollar's monopoly by a small increment. Enough small increments and the monopoly erodes.

The dollar's reserve currency status gives the United States an enormous structural advantage: the ability to borrow in its own currency at low interest rates, to run trade deficits that would cripple other countries, and to use financial sanctions as a weapon of foreign policy. If that advantage erodes, the fiscal foundation of American power erodes with it. China doesn't need to kill the dollar. It just needs to wound it enough that the US can't use it as a weapon anymore.

• • •

The sum of all these pieces, the infrastructure, the resource access, the energy transition dominance, the financial architecture, is an empire that is being assembled in plain sight while the United States looks the other way.

It's not invisible. Analysts write about it. Think tanks publish reports. Journalists investigate individual projects. The information is all there. What's missing is the response. The institutional framework to counter Chinese economic statecraft doesn't exist because the institution that makes strategy in America is the Pentagon, and the Pentagon makes military strategy, and military strategy has no tool for this.

You can't sink a Belt and Road loan with a Tomahawk missile. You can't blockade a rare earth processing plant. You can't scramble F-35s to intercept a fiber optic cable. The entire American strategic apparatus is designed to win wars, and China has built an empire without starting one.

That's the genius of the approach. And the grand bargain for the Russian Far East is the move that completes it. If that bargain happens, it gives China the one thing it doesn't currently have: resource self-sufficiency. And a resource-self-sufficient China that controls the energy transition supply chain, the Eurasian infrastructure network, and an alternative financial system is a China that doesn't need the rest of the world's permission for anything.

Not an invasion of Taiwan. A continent-spanning economic empire that makes China untouchable.

Unless someone notices in time to do something about it. So far, nobody has.

Chapter 19: The Grand Bargain

The Russian Far East is a territory so vast it defies intuition. Stretch it across a map and it covers an area larger than the European Union. From the Ural Mountains to the Pacific coast, from the Mongolian border to the Arctic Ocean, it encompasses taiga forests, mountain ranges, river systems, mineral deposits, and coastline that would make it one of the richest territories on Earth if anyone could figure out how to develop it.

Almost nobody lives there. The entire Russian Far East Federal District has a population of roughly 8 million people, down more than 10 percent in two decades, and that number has been shrinking steadily. Young Russians leave for Moscow or St. Petersburg, where the jobs, the culture, and the future are. The villages that dot the territory are aging and emptying. The infrastructure, most of it built during the Soviet era, is crumbling. The roads are often unpaved. The railroads are slow. The cities feel like time capsules from the Brezhnev era, except colder and more decrepit.

On the Chinese side of the border, in Heilongjiang province and the surrounding northeast, live over 100 million people. The demographic disparity is staggering: more than fifteen Chinese citizens for every Russian one, separated by a river border and a set of historical grievances that Moscow tries not to think about.

Russia has had 400 years to develop the Far East. It has consistently failed. The territory was conquered by Cossack explorers in the 17th century, exploited for furs, defended against Japanese encroachment in the early 20th century, industrialized crudely under Stalin, neglected under Brezhnev, and abandoned economically after the Soviet collapse. Moscow has launched development initiative after development initiative, announced grand plans, built showcase projects, and watched them all decay into irrelevance because the fundamental problem is unsolvable: you cannot develop a territory the size of a continent with 8 million people and a government 4,000 miles away that doesn't care about it.

• • •

Now look at Russia's position after Ukraine.

The professional military that Putin spent twenty years rebuilding has been shattered. By the end of 2025, Russia had suffered an estimated 1.2 million battlefield casualties, including roughly 300,000 killed, according to a CSIS analysis---more losses than any major power has sustained in any conflict since World War II. Many of Russia's best-trained officers and special operations forces are dead or permanently disabled. The equipment losses, thousands of tanks, armored vehicles, aircraft, and helicopters, have gutted the conventional military's capacity to fight a second war anywhere. The vaunted Russian defense industry, which was supposed to outproduce Ukraine's Western-supplied arsenal, has struggled to keep up with demand and has been forced to buy artillery shells from North Korea and drones from Iran.

The economy is a Chinese dependency in all but name. Western sanctions cut Russia off from most of the global financial system, forcing it to redirect trade flows eastward. China is now Russia's largest trading partner by a wide margin. Russian energy exports, which used to flow west to Europe, increasingly flow east to China through pipelines that were built to serve the Chinese market. The Power of Siberia pipeline, which began operating in 2019, delivers Russian gas directly to northeastern China and reached its design capacity of 38 billion cubic meters per year in 2025.

The Power of Siberia 2, a proposed 2,600-kilometer pipeline that would carry 50 billion cubic meters annually from Russia's Yamal gas fields to China via Mongolia, has been under negotiation for years. In September 2025, Gazprom announced a "legally binding memorandum" with China, but the key terms---price, investment structure, and delivery start date---remain unresolved. China has been slow-walking the deal deliberately, using it as leverage in broader energy negotiations while Russia grows more desperate. Even if a final contract is signed in 2026, construction would take roughly five years, and another five to reach full capacity. The pipeline would not meaningfully boost Russian export revenues until the mid-2030s.

Beijing is in no hurry. Moscow is.

Russian imports of Chinese manufactured goods have surged to replace Western products lost to sanctions. Chinese cars are flooding the Russian market. Chinese electronics fill the stores. Chinese telecommunications equipment is replacing the Ericsson and Nokia systems that were pulled out. The technological dependency is deepening with every passing month, and there is no alternative supplier. Europe won't trade with Russia. America won't. Japan won't. China is the only game in town, and China knows it.

Putin thinks he's playing a partnership of equals. He's selling the furniture to pay the rent.

* * *

The deal, when it comes, won't look like a deal. That's the important thing to understand. There will be no announcement that China is acquiring Russian territory. No treaty. No map with new borders. The word "cession" will never appear in any document. Instead, there will be a series of agreements that, individually, look like normal bilateral cooperation and, collectively, transfer effective control of the Russian Far East to China over a period of years.

Long-term sovereign leases on resource extraction zones. Fifty-year agreements. Ninety-nine-year agreements. Joint development zones where Chinese companies invest, Chinese workers operate, and Chinese security provides protection. Special economic zones governed by rules that effectively replicate Chinese commercial law within Russian territory. Agricultural leases on vast tracts of land that Russian farmers aren't using because there aren't enough Russian farmers.

Chinese-built infrastructure connecting the territory to China rather than to Moscow. Railroads running south to the Chinese border, not west to the Urals. Pipelines flowing east. Telecommunications networks integrated with Chinese systems. Power grids tied to Chinese generation. Each project makes perfect commercial sense. Each project deepens the dependency. And each project makes the territory more Chinese in function even as it remains Russian on the map.

The population shift will follow the infrastructure. Chinese workers come to staff the projects. Chinese managers come to run the operations. Chinese families come because the jobs are there. Chinese businesses open to serve the Chinese workers. Chinese schools open for the Chinese children. Over a decade or two, the demographic balance that is already fifteen to one tips further. The Far Eastern cities that are emptying of Russians fill with Chinese residents who are there legally, employed productively, and integrated into an economic system that runs on Chinese capital.

Russia doesn't lose the territory. It just stops mattering that Russia technically owns it.

• • •

The leverage comes from Ukraine. That's where the bargain's terms get set.

Russia needs an end to the Ukraine war that lets Putin, or whoever succeeds him, survive politically. A settlement that allows Russia to keep enough territory to claim victory. Diplomatic support that prevents the international community from treating Russia as a pariah indefinitely. Economic support to rebuild a country that has been under comprehensive sanctions for four years. A financial system to operate through when the Western banking system remains hostile.

China can provide all of that. Beijing can pressure European governments to accept a settlement by threatening economic consequences for non-cooperation. It can provide the diplomatic cover at the United Nations that Russia needs to re-enter the international system. It can provide the economic lifeline, the investment capital, the manufactured goods, the technology transfers, that Russia needs to rebuild. And it can provide the financial infrastructure through CIPS and yuan-denominated trade that lets Russia function economically outside the Western system.

The price is the Far East.

Not stated that baldly, of course. Nobody puts it in those terms. The price is "deepened cooperation" and "strategic partnership" and "joint development" and "mutually beneficial resource agreements." The words sound like diplomacy. The substance is transfer.

• • •

Putin can't refuse. That's the structural reality that makes the bargain inevitable rather than speculative.

Russia can't develop the Far East on its own. It has tried for 400 years and failed. It doesn't have the population, the capital, or the institutional capacity. The territory is an asset on the balance sheet that produces a fraction of its potential value because Russia can't unlock it.

Russia can't defend the Far East against China if the relationship sours. The force disparity is absurd even before Ukraine. After Ukraine, with the conventional military hollowed out and the nuclear deterrent being the only credible defense, it's even more lopsided. China is not going to invade. It doesn't need to. But the implicit understanding that China could apply pressure if cooperation faltered is part of the negotiating dynamic, even if nobody says it out loud.

Russia can't pivot back to the West. The bridges are burned. European public opinion on Russia is toxic. American sanctions aren't going to be lifted comprehensively regardless of who is president. Japan has territorial disputes with Russia over the Kuril Islands. There is no Western partner waiting to offer an alternative to Chinese economic integration. China is the only option, and when you only have one option, you take what's offered and you don't negotiate too hard on the terms.

And China knows all of this. Xi Jinping is not sentimental about partnerships. He's transactional. Russia provides what China needs: energy, resources, a buffer against Western encirclement, and diplomatic support. In exchange, Russia gets what it needs to survive. But the terms of exchange are set by the party with options, not the party without them. And China has options. Russia doesn't.

• • •

Then the soldiers come.

Not as invaders. As protectors. The same template China has used in the South China Sea, applied to a different geography.

In the South China Sea, China didn't invade islands. It built them. Artificial land on submerged reefs. Then research stations on the artificial land. Then airstrips on the research stations. Then radar installations on the airstrips. Then missile batteries next to the radar. Each step was too small to trigger a military response from any single claimant. Each step was irreversible. And now China has de facto military control over one of the most important waterways on the planet, and nobody is going to make them leave because the cost of doing so exceeds the cost of tolerating it.

Apply that template to the Russian Far East. Chinese-built ports on the Arctic coast, maintained by Chinese workers, protected by Chinese private security. Chinese-built rail junctions with Chinese-operated loading facilities and Chinese-employed maintenance crews. Chinese-built resource extraction operations with Chinese-operated heavy equipment and Chinese-managed logistics. At each facility, the security presence is minimal at first. A few guards. A small company. Then a garrison. Then a permanent base to safeguard the cooperative development zone.

The Russian authorities don't object because the Chinese presence is the only thing keeping the economy of the region functional. The local population, what's left of it, doesn't object because the Chinese investments are the only source of jobs. Moscow doesn't object because the alternative to Chinese development is no development at all, and the revenues from the joint ventures are funding the government.

And at some point, probably within 15 to 20 years of the initial agreements, the Far East is Chinese in everything but the flag. The population is substantially Chinese. The economy is entirely Chinese-dependent. The infrastructure connects to China, not to Moscow. The security is provided by Chinese forces. The governance is nominally Russian but functionally Chinese-directed because every important economic decision runs through Chinese-controlled institutions.

That's not invasion. That's absorption. And it's how empires have been built since the beginning of recorded history. Not through conquest, though there's plenty of that in the record. Through making yourself indispensable to someone who's desperate, and then collecting the bill when they can't pay.

* * *

The historical irony is almost too perfect. Russia's entire strategic identity for three centuries has been about territorial expansion. Pushing east across Siberia. Pushing south into Central Asia. Pushing west into Europe. The Russian empire was built by a civilization that defined itself through the acquisition and defense of territory more than any other great power in modern history.

And the end result of Putin's grand adventure in Ukraine, the war that was supposed to restore Russian greatness and push NATO back from Russia's borders, may be the loss of everything east of the Urals. Not to conquest. To a handshake. To a deal made by a leader who had no other options, with a partner who had all of them.

The territory that Russia spent 400 years conquering and couldn't develop. The resources that Russia sat on and couldn't exploit. The coastline that Russia controlled and couldn't utilize. All of it transferred, over the course of a generation, to a neighbor with the population, the capital, the technology, and the patience to do what Russia never could.

China doesn't need to fight Russia. It just needs to wait for Russia to need China badly enough to sell the one thing it has left that's worth buying.

* * *

What does this mean for the rest of the world?

A China that controls the Russian Far East, in addition to its existing Belt and Road network, is a China that has solved its resource problem. Energy from Russian gas and oil, delivered by pipeline rather than through vulnerable sea lanes. Minerals from Siberian deposits. Timber from the world's largest forests. Freshwater from Lake Baikal and the river systems of eastern Siberia. Agricultural land that's becoming more productive as the climate warms. Arctic coast access for the Northern Sea Route. Everything a civilization needs to sustain itself, independent of any supply chain that the US Navy could interdict.

That's the endgame that Chapter 17 described. Resource self-sufficiency. The Belt and Road provides the global network. The Russian Far East provides the resource base. Together, they give China the ability to function as an autarkic empire if it has to, while maintaining global commercial engagement when it wants to. The Strait of Malacca, which is currently China's version of the Hormuz vulnerability, becomes irrelevant if energy and resources arrive overland from the north.

For the United States, this scenario presents a strategic challenge that is different in kind from anything in its experience. American strategy for the past eighty years has been based on the ability to use naval power and economic sanctions to constrain adversaries. Blockade their ports. Interdict their supply lines. Freeze their assets. Cut them off from global markets. A China that can't be blockaded, can't be sanctioned effectively, and can't be cut off from essential resources is a China that the current American strategic toolkit cannot handle.

That doesn't mean the US is powerless. But it means the tools have to change. And that's what Part Six is about. The tools we're not using. The tools that could rebuild American capacity and create a different kind of competitive advantage. Starting with the financial extraction that's hollowed out the institutions we need, and then moving to the technologies that could change everything.

The quiet empire is being built. The grand bargain is coming. The question is whether the country that should be responding has the capacity and the will to respond with something other than aircraft carriers and Tomahawk missiles.

So far, the answer is no. But it doesn't have to stay that way.

Part Six begins with why the answer is no, and then shows how it could become yes.

Part Six: The Hollowed-Out Country and the Tools to Rebuild It

Chapter 20: The Enshittification of Strategic Capacity

I wrote an entire book about this. It's called *The Enshittification of America: How Private Equity Destroyed the Things We Love*, and it documents, industry by industry, how the leveraged buyout model has gutted twelve sectors of the American economy, from airlines to hospitals to newspapers to grocery stores. If you want the full autopsy, the detailed case studies, the named firms and the named executives and the specific mechanisms of destruction, that book is where to find it.

This chapter is different. This chapter is about what the destruction means for the country's ability to survive what's coming. Because every crisis described in the first five parts, every military challenge, every geopolitical competition, every structural force, requires institutional capacity to respond to. Factories that can build things. Hospitals that can treat surges. Media that can inform the public. A workforce that can show up and perform. An industrial base that can scale when it's needed.

Yes, it angers me. The same pattern that gutted airlines, hospitals, and grocery stores showed up in the defense industrial base and nobody stopped it. Private equity extracted value from shipyards, from ammunition manufacturers, from the suppliers that make the components no one thinks about until they are missing. Quarterly returns were optimized. Long-term capacity was destroyed. And now we are in a position where we cannot build ships fast enough, cannot replenish ammunition fast enough, and cannot surge production when we need to because the industrial capacity to do so was sold off a decade ago to fund someone's carried interest. The people who did this are not in jail. They are at conferences.

That capacity has been systematically hollowed out. Not by foreign competitors. Not by inevitable market forces. By a specific financial model, operated by specific firms, enabled by specific policy choices, that extracts wealth from productive institutions and leaves the wreckage for everyone else to deal with.

• • •

The model is simple enough to explain in a paragraph. A private equity firm identifies a profitable business. It acquires the business using mostly borrowed money. It loads the acquisition debt onto the acquired company, not onto the PE firm. Then it strips costs: fires experienced workers, defers maintenance, cuts research and development, sells real estate, outsources everything that can be outsourced. It extracts management fees and dividends regardless of whether the business is profitable. Then it either flips the company to the next buyer at a markup, having dressed up the short-term financials by cutting everything that matters for the long term, or it lets the company collapse, having already pulled out more than it paid. The debt stays with the corpse.

This is not capitalism. Capitalism is supposed to allocate resources to their most productive uses. This is extraction. It allocates resources to the already-wealthy by stripping them from productive enterprises that serve communities, employees, and customers. The difference matters, and people who defend private equity by calling criticism "anti-capitalist" are either confused about what capitalism means or deliberately obscuring what PE does.

• • •

Now connect that model to the strategic capacity America needs.

The defense industrial base. Chapter 2 described the production bottleneck: the US can't build missiles fast enough because the sub-tier suppliers that make the components don't have the capacity. Many of those suppliers have been through private equity ownership cycles. TransDigm Group is the poster child. Founded in 1993 by a private equity firm, passed through two more PE owners before going public, TransDigm has acquired over 90 aerospace component companies. It operates like a PE firm wearing an aerospace mask: buy a sole-source supplier of a niche aircraft part, use the monopoly position to jack up prices, and extract returns that would make a hedge fund blush.

A Pentagon Inspector General investigation found TransDigm earning profit margins of up to 4,436 percent on individual parts, including a case where a component that should have cost $32 was sold to the Department of Defense for $1,443. The company supplies parts for nearly every military aircraft in the fleet. It is, in a sense, too embedded to punish and too profitable to reform.

This pattern is identical to every other industry: buy the company, strip the costs, extract the fees, sell or abandon the shell. The experienced workers get fired because labor is the biggest cost line. The R&D budget gets cut because R&D doesn't produce returns this quarter. The maintenance gets deferred because maintenance is a cost, not a revenue line. The redundant capacity, the extra tooling, the backup systems, the reserve workforce, gets eliminated because redundancy is inefficiency and inefficiency is the enemy of returns.

When the war comes and the Pentagon throws money at production, the money hits organizations that have been optimized for peacetime extraction rather than wartime surge. TransDigm is the most visible example, but it is not an outlier. Private equity firms have acquired over 500 defense contractors since the early 2000s, according to the Journal of Global Security Studies. In April 2024, Arcline Investment Management took Kaman Corporation private in a $1.8 billion deal. Kaman, founded in 1945, makes aircraft bearings, seals, composite aerostructures, and helicopter components for nearly every platform in the military fleet. Once private, its financial data vanished.

The Stockholm International Peace Research Institute warned that PE acquisitions are making it impossible to track what is happening inside the defense supply chain. Companies "disappear into a black box" after acquisition, SIPRI researchers said.

The people are gone. The institutional knowledge left with them. The tooling is at capacity with no margin. The suppliers of specialty components are sole-source because the competitors were consolidated or closed during previous PE cycles. The bottleneck isn't funding. It's the human and physical capital that was extracted over the preceding decade by firms that are already counting their returns from the next deal.

The shipyard workforce problem from Chapter 3 has PE fingerprints on it. The defense shipbuilding industry consolidated not purely because of market forces but because financial engineering made consolidation profitable for the firms doing the consolidating. Yards got bought, merged, stripped of excess capacity, and rationalized. The workforce fluctuated through cycles of layoffs and hiring driven by financial optimization rather than production needs. The skilled tradespeople who got laid off during a down cycle didn't sit around waiting for the up cycle. They moved on, took their expertise with them, and often left the industry permanently.

• • •

Healthcare. Chapter 16 described the pandemic vulnerability: a healthcare system that can't handle a surge because it's been optimized to run at maximum capacity with minimum slack. Private equity is a primary reason why.

PE firms have been buying hospitals, nursing homes, physician practices, and emergency staffing companies at an accelerating rate for fifteen years. The pattern in healthcare is identical to every other industry: buy, load with debt, cut staff, reduce quality, extract fees.

Steward Health Care is the case study. In 2010, private equity firm Cerberus Capital Management bought a struggling Catholic hospital system in Massachusetts called Caritas Christi for roughly $900 million, renamed it Steward, and promised the state it would invest in the facilities. Instead, Cerberus sold the hospital real estate to a landlord called Medical Properties Trust for $1.25 billion, burdening Steward with hundreds of millions in annual lease payments. The CEO, Ralph de la Torre, paid himself and the ownership a $111 million dividend and bought a $40 million yacht. By 2024, Steward had grown to 31 hospitals across eight states and then filed for bankruptcy. Two Massachusetts hospitals, Carney Hospital in Boston's low-income Dorchester neighborhood and Nashoba Valley Medical Center in central Massachusetts, closed permanently. Over 150,000 patients lost their local healthcare access. Cerberus walked away with an estimated $800 million in profit. De la Torre was subpoenaed by Congress and refused to show up.

212

Emergency rooms are staffed by PE-owned companies that optimize for patient throughput rather than patient care. Nursing homes under PE ownership had significantly higher COVID death rates, not because the virus was different in those facilities but because the staffing had been cut so deep that there weren't enough hands to care for the patients. A 2021 NBER study by researchers at the University of Pennsylvania, NYU, and the University of Chicago found that PE-owned nursing homes had a 10 percent higher mortality rate for Medicare patients. Over twelve years, that translates to roughly 20,000 excess deaths. The mechanism was simple: after a PE acquisition, frontline nursing staff hours dropped by 3 percent while the use of antipsychotic drugs to sedate patients jumped by 50 percent. Fewer nurses, more drugs, more dead patients, higher Medicare bills.

Rural hospitals are closing across America, and PE extraction is a significant contributor. A hospital gets acquired. The profitable service lines get kept. The unprofitable ones, which are often the ones the community depends on most, get cut. The real estate gets sold. The debt load makes the operation unviable. The hospital closes. The community loses its healthcare access. The PE firm has already extracted its returns and moved on.

When the next pandemic hits, it hits a healthcare infrastructure that is weaker, more fragile, and less capable of surging than what we had for COVID. The beds that were eliminated for efficiency can't be put back in an afternoon. The nurses who were laid off didn't wait around. The supply chains for critical drugs and equipment that were streamlined for just-in-time delivery can't handle demand spikes. Everything that the PE model optimized for, efficiency, low inventory, minimal staffing, maximum throughput, is exactly the opposite of what a pandemic response requires.

• • •

Media. Chapter 5 described the "wars without reasons" problem: 65 percent of Americans couldn't understand the goals of the Iran war. Part of the reason is that the journalism that would explain it has been killed.

Alden Global Capital, the hedge fund that is the single most destructive force in American journalism, has been buying newspaper chains for the explicit purpose of extracting revenue while the print business dies. In 2021, Alden bought Tribune Publishing for $630 million, acquiring the Chicago Tribune, the Baltimore Sun, the New York Daily News, the Hartford Courant, and the Orlando Sentinel. The Chicago Tribune lost a quarter of its newsroom within months. The Denver Post, which Alden already owned through its Digital First Media chain, had been cut by a third.

Between 2012 and 2019, according to the NewsGuild, Alden cut 71 percent of jobs across its unionized newsrooms. Gregory Pratt, the Chicago Tribune's City Hall reporter and head of its newsroom union, told NPR what everyone in the industry already knew: "They're shameless about cutting. They don't care about the social value of the news." Alden's president, Heath Freeman, insisted the company was finding a "sustainable path" for local news. What he meant was sustainable extraction.

Buy the chain. Slash the newsroom. Sell the real estate. Defer maintenance on the printing presses. Extract advertising revenue and subscription income from the remaining skeleton staff. When the paper finally folds, the real estate sale has already paid for the acquisition and generated a return. The journalism was never the point. The cash flow was the point. The journalism was just the machine that generated the cash flow, and like every machine in a PE portfolio, it gets run until it breaks and then it gets abandoned.

The collapse of local journalism means communities lose the information infrastructure that democracy depends on. City councils meet without reporters present. School boards make decisions nobody covers. Police departments operate without oversight. Corruption goes unreported because there's nobody to report it. And at the national level, the public can't understand why their country is at war because the newsrooms that would investigate, contextualize, and explain the conflict have been hollowed out by financial extraction.

A democracy that can't inform its citizens can't hold its leaders accountable. A public that doesn't understand what its government is doing can't make rational decisions about whether to support or oppose it. The dysfunction described in Part Two, wars without reasons, a public that doesn't understand why it's fighting, isn't just a failure of political leadership. It's a failure of the information system that's supposed to connect leaders to citizens. And that information system was killed, in significant part, by private equity.

• • •

Housing. The anti-immigration politics that threaten the demographic advantage described in Chapter 12 are driven in significant part by housing costs. And housing costs are driven in significant part by private equity.

After the 2008 financial crisis, PE firms and institutional investment vehicles bought hundreds of thousands of single-family homes at foreclosure prices. Blackstone Group led the charge, spending over $10 billion through its subsidiary Invitation Homes to acquire roughly 80,000 single-family houses across markets like Atlanta, Phoenix, Las Vegas, and Tampa, places where the foreclosure crisis had crushed prices. Invitation Homes became the largest single-family rental company in the country, went public in 2017, and kept buying. In some Sunbelt markets, institutional investors own 20 to 30 percent of the single-family rental stock. Families who lost their homes in the crisis found themselves renting from the firms that bought those homes at pennies on the dollar. Rents went up. Maintenance complaints went up. Homeownership rates went down. Communities that were already devastated by the financial crisis became extraction zones for Wall Street capital. The model spread. American Homes 4 Rent, Progress Residential, and others piled in. By the mid-2020s, institutional investors owned an estimated 600,000 to 700,000 single-family homes across the country, and the number keeps growing.

The housing affordability crisis is one of the primary drivers of the political anger that fuels anti-immigration sentiment. People who can't afford houses look for someone to blame, and immigrants are an easier target than Wall Street firms that most people have never heard of. The PE extraction of the housing market doesn't just transfer wealth upward. It poisons the political environment in ways that undermine the strategic interests this book identifies as critical.

• • •

The China contrast is the part that should make every American angry.

When Beijing decides to build strategic capacity, the capital goes to building capacity. When China invests in semiconductor fabrication, the money goes to fabs. When China builds nuclear reactors, the money goes to reactors. When China funds rare earth processing, the money goes to processing plants. The state directs investment toward strategic objectives, and the investment arrives at its destination because there is no intermediary whose business model is to extract a portion of every dollar that flows through the system.

This doesn't mean the Chinese model is better in all respects. It produces its own pathologies, and they are severe.

The property sector that drove 25 percent of GDP has imploded. Evergrande, once the world's most indebted developer, was liquidated with over $300 billion in unpaid debts and hundreds of thousands of unfinished apartments. Country Garden, the second-largest developer, defaulted in 2023. By 2025, new residential development had fallen 58 percent from 2019 levels and Goldman Sachs estimated the total value of unsold and unfinished homes at $4.1 trillion. Youth unemployment hit 20 percent before Beijing stopped publishing the number. Local governments are buried under an estimated $7 to $10 trillion in hidden debt from off-balance-sheet financing vehicles used to fund infrastructure during the boom years.

The social contract that sustained CCP legitimacy for forty years, shut up and get rich, is breaking because people stopped getting rich. A generation of Chinese homebuyers who poured their savings into apartments that were never built are not grateful citizens of a rising empire. They are angry people whose wealth evaporated while the system that promised them prosperity arrested anyone who complained about it.

The pattern should look familiar. Chinese developers did to homebuyers what American PE firms did to hospital patients and newspaper readers: extract the value, load the debt onto the institution, and walk away before the collapse. The mechanism differs. The outcome, ordinary people bearing the losses of a system designed to enrich the operators, is identical. But in the specific context of building strategic industrial capacity, a system that protects long-term investment beats a system that strips it. The PE model is a uniquely American pathology, shared to some extent by the UK, that creates a structural disadvantage in any long-term competition against an economy that directs capital rather than allowing it to be extracted. China's economy is fragile. Its strategic positioning is not.

The US is trying to compete with China in semiconductors, in clean energy manufacturing, in shipbuilding, in missile production, with an economic system that siphons a percentage of every investment dollar into management fees, leveraged dividends, and financial engineering. It's like trying to fill a bathtub with someone drilling holes in the bottom. The water goes in at the top, the PE firms drain it from below, and the tub never fills.

• • •

The optimistic note, because this book always has one, is that every element of the PE extraction model is a policy choice. And policy choices can be changed.

The carried interest loophole, which allows PE managers to pay capital gains tax rates instead of income tax rates on their fees, could be closed with a single act of Congress. The tax deductibility of debt interest, which makes leveraged buyouts financially attractive by letting the acquired company deduct the interest on its own acquisition debt, could be reformed. Bankruptcy protections that allow PE firms to walk away from the consequences of their decisions while retaining the profits extracted before the collapse could be tightened. Regulatory oversight of PE acquisitions in strategic industries, defense suppliers, healthcare, media, housing, could be strengthened.

None of these reforms would end private equity. PE, at its best, provides capital to underperforming companies, installs better management, and creates value that benefits everyone: owners, workers, customers. That version of the industry exists. It is not the version described in this chapter. The reforms would end the most predatory version of it, the version that treats productive institutions as carcasses to be stripped, and redirect the industry toward the value creation that PE firms advertise in their pitch decks but rarely deliver in practice.

218

The political obstacle is real. The PE industry spends hundreds of millions on lobbying and campaign contributions. The revolving door between Wall Street, the Treasury Department, and the SEC ensures that the regulators share the worldview of the regulated. The intellectual infrastructure of free-market economics provides academic cover for extraction by calling it "efficiency." And the complexity of the financial instruments involved means that most voters, and most legislators, don't understand what PE does until the hospital closes or the newspaper folds or the missile factory can't ramp production.

But the Iran war is making the cost visible in a way that financial abstractions never do. When the country can't build missiles fast enough to fight a war, and the reason it can't is that the supply chain was hollowed out by financial engineering, the connection between Wall Street extraction and national security becomes concrete enough for even a Congress member to understand.

Whether they'll act on that understanding is another question. But at least the question is being asked. And this book, along with *The Enshittification of America*, is part of asking it.

Chapter 21: The Deskilled Nation

Every solution in this book requires someone to build it.

Nuclear reactors require engineers who understand reactor physics, radiation health physicists who can design safety systems, welders certified to nuclear standards, electricians who can wire control systems to tolerances that leave no room for error. AI systems require mathematicians, computer scientists, data engineers, and the specialists who train and maintain the models. Longevity research requires molecular biologists, geneticists, clinical researchers, and the lab technicians who keep the experiments running. Human enhancement technology requires neuroscientists, biomedical engineers, surgeons, and the manufacturing workforce that builds the devices. Ships require welders, pipefitters, electricians, and riggers. Missiles require rocket motor technicians, seeker head assemblers, and quality control inspectors who can tell by touch when something isn't right.

The United States educational system is producing fewer of all of these people every year. And it is also degrading the baseline competence of the general population to a degree that threatens the country's ability to function as an advanced technological civilization.

That last sentence sounds alarmist. It isn't. Numbers are worse than most people realize because most people haven't looked at them.

• • •

American 15-year-olds rank 26th in the world in mathematics on the Programme for International Student Assessment, the PISA tests that compare educational outcomes across developed countries. Twenty-sixth. Behind

Estonia. Behind Slovenia. Behind the Czech Republic. Behind countries with a fraction of America's wealth, a fraction of its per-pupil spending, and none of its advantages in university research infrastructure. The science ranking is better---10th---but the math result is the one that matters for the technical workforce this book keeps describing as critical.

More than half of American adults cannot read at a sixth-grade level. That's not a misprint. According to the Barbara Bush Foundation's analysis of federal assessment data, roughly 130 million adults in the United States, 54 percent of those between the ages of 16 and 74, lack the literacy skills to read a newspaper article, follow written instructions for a medication, or understand a lease agreement. They are functionally illiterate in a society that runs on written information.

The national security implication of that number is not abstract. Chapter 11 described the information warfare campaigns that Russia, China, and Iran are running against democratic societies. Those campaigns succeed when populations cannot evaluate the quality of what they read. A citizen who can parse a news article, identify a source, and distinguish an argument from an assertion has some natural resistance to manipulation. A citizen who reads below a sixth-grade level does not. Societies with high levels of trust, strong institutions, and resilient media ecosystems resist information warfare better than fragmented ones. America is not that society. It is a country where 130 million adults cannot read well enough to follow a lease agreement, where social media algorithms feed them content calibrated to provoke rather than inform, and where the education system that was supposed to produce informed citizens has instead produced a population uniquely vulnerable to the cheapest weapon in any adversary's arsenal. Literacy is not just an economic issue or an education issue. It is a defense issue.

The average American college graduate today demonstrates roughly the literacy competency of a high school graduate from 1970. Grade inflation has made the credential meaningless as a signal of ability. An A in 1970 meant mastery of the material. An A in 2025 means you showed up and didn't cause trouble. The devaluation of the degree has been so thorough that employers increasingly ignore educational credentials entirely and test for competency directly, which tells you everything you need to know about what the education system is producing.

STEM degree production among domestic American students has been flat for years. The gap between what the economy needs and what the universities produce has been filled by international students, many of whom return to their home countries after graduating. The US produces roughly 400,000 STEM graduates a year. China produces 3.5 million. Even accounting for differences in quality and methodology, the volume gap is enormous, and volume matters when you're trying to build an industrial base.

• • •

The K-12 system is where the damage starts, and it's where the most contentious arguments live. This book is going to avoid the culture war and focus on what the numbers say.

American public schools spend more per pupil than almost any country on Earth. The national average exceeds $16,500 per student per year. New York spends over $35,000. New Jersey tops $30,000. These are extraordinary sums by any international standard. And the outcomes, as measured by every available metric, are mediocre to terrible. New York City illustrates the gap with painful clarity. The nation's largest school district spends over $32,000 per student per year and employs more administrators per pupil than almost any district in the country. On the 2024 National Assessment of Educational Progress, only 28 percent of NYC fourth graders were proficient in reading, below the national average of 31 percent. Only 23 percent of eighth graders were proficient in math, well below the national rate of 28 percent. Forty percent of the nation's fourth graders are now working below the NAEP basic level in reading, the highest percentage since 2002. The money is being spent. The results are not being produced.

The money isn't reaching the classroom. Administrative bloat consumes a growing share of school budgets. The ratio of administrators to teachers has exploded over the past forty years. Compliance costs, driven by federal and state mandates, consume time and resources that should go to instruction. Professional development programs pull teachers out of classrooms to attend workshops of dubious value. The bureaucratic overhead of running a school district in the United States is staggering, and almost none of it makes students better at mathematics.

The teaching profession itself has been degraded to the point where the talent pipeline is collapsing. Starting salaries for teachers are below starting salaries for almost every other profession that requires a four-year degree. In many states, they're below the median income. Who wants to spend four years getting a degree, take on student debt, and then earn less than a first-year plumber? The people who used to become teachers, smart, dedicated, civic-minded college graduates, now become software engineers, consultants, or anything else that pays a living wage. The schools get whoever is left.

This is not the teachers' fault. The teachers who remain in the profession are often heroic, doing difficult work under impossible conditions for insufficient compensation. The fault belongs to a system that decided education was a cost center to be minimized rather than an investment to be maximized, and that applied the same extraction logic described in Chapter 20 to human capital development: strip the inputs, defer the maintenance, optimize the metrics, and hope that nobody notices the output is getting worse.

• • •

Higher education is a different failure, but it connects to the same structural pattern.

The American university system was, for most of the 20th century, genuinely the best in the world. It produced more Nobel laureates, more breakthrough research, more technological innovation, and more educated citizens than any other system. It was the engine of American competitive advantage, the place where the ideas and the people that drove the postwar economy were forged.

It has been financialized into a debt extraction machine.

Student debt in the United States exceeds $1.7 trillion. The average bachelor's degree recipient carries roughly $30,000 in loans. Many carry far more. This debt was taken on in exchange for degrees of declining value from institutions that have optimized for enrollment revenue rather than educational quality. Tuition has increased at roughly three times the rate of inflation for four decades, driven not by improved instruction but by administrative expansion, amenity competition, and the simple fact that government-backed student loans made it possible to charge whatever the market would bear because the students weren't paying with their own money. They were paying with their future earnings, pledged in advance to loan servicers.

The for-profit education sector was substantially a private equity play. Companies like the University of Phoenix and DeVry were designed to maximize federal student loan extraction. They targeted veterans using GI Bill benefits and low-income students using Pell Grants and federal loans, produced credentials of minimal market value, and left the students with debt and no meaningful improvement in their employment prospects. Billions in federal education funding were siphoned into PE returns rather than education. The students got a piece of paper. The PE firms got the money.

Meanwhile, the research university system, still world-class at the very top, has become increasingly dependent on foreign graduate students and postdoctoral researchers. Walk through any top-tier STEM department and count the nationalities. The research that maintains American technological leadership is being done, in significant measure, by people who weren't educated in American K-12 schools. They were educated in Chinese, Indian, Korean, and European schools, came to America for graduate school because American universities are still the best place to do research, and many of them now go home or to third countries after completing their degrees because American immigration policy makes it difficult for them to stay.

The talent pipeline that feeds the innovation ecosystem is leaking from both ends. Domestic students aren't coming out of K-12 prepared for rigorous STEM education. Foreign students who fill the gap face immigration barriers that encourage them to take their American-trained expertise elsewhere. The pipeline still produces enough exceptional talent to keep the system running. For now. But the margins are thinner than they've ever been, and the trajectory is going the wrong direction.

• • •

The vocational and technical education collapse is the part that connects most directly to the military and industrial capacity problems described in Part One.

The United States dismantled its vocational training system over the past forty years. The decision, never made explicitly but enacted through a thousand budget cuts and policy changes, was that everyone should go to college. Vocational education was stigmatized as a track for students who weren't smart enough for "real" school. Shop classes were eliminated. Apprenticeship programs were defunded. Community colleges shifted their emphasis from technical training to transfer preparation. The entire infrastructure that produced skilled tradespeople, the welders, machinists, electricians, plumbers, HVAC technicians, and industrial workers who build and maintain the physical systems that civilization depends on, was allowed to wither.

The consequences are visible in every sector. Roughly 62 percent of American adults do not have a bachelor's degree. The system that was supposed to serve them was gutted. The jobs that were supposed to replace the ones they lost to offshoring and automation were supposed to come from retraining programs that largely failed. The opioid crisis, the deaths of despair, the political rage that powered populist movements on both the left and the right: all of it traces back, at least in part, to a generation of Americans who were told that the old economy was over and the new one would provide, and then discovered that the new economy had no use for them and no interest in teaching them how to fit in. A welder who loses a factory job at 45 cannot become a software engineer through a six-week coding bootcamp, regardless of what the promotional materials say. The skills gap is not a training gap. It is a structural mismatch between what the economy demands and what the education system produces, and it has been growing for forty years.

Germany took a different path and the results are instructive. The German dual education system, in which roughly half of all young people enter apprenticeship programs that combine classroom instruction with on-the-job training, produces a skilled industrial workforce that is the envy of the developed world. German manufacturing workers are among the most productive on the planet. German shipyards build ships on time. German factories produce precision machinery. The vocational training system is the foundation of the German economic model, and it is treated with the same seriousness that America reserves for its university system.

The US shipyard workforce crisis from Chapter 3 is a vocational education failure. You cannot build a nuclear submarine with college graduates who studied communications. You need welders certified to nuclear standards, and those welders require five to seven years of specialized training that no university offers. You need pipefitters who understand the tolerances of high-pressure steam systems. You need electricians who can wire control systems for nuclear reactors. None of these skills are taught in the educational system that America decided was sufficient for its needs.

The result: four shipyards, all behind schedule, all struggling to hire qualified workers, building ships for a navy that is too small for its strategy, in a country that decided forty years ago that tradespeople were less important than knowledge workers and is now discovering that knowledge workers can't weld.

• • •

This isn't just a workforce problem. It's the institutional manifestation of the cognitive decline that *The Death of Thinking*, the first book in this series, diagnosed at the individual level. A population that has lost the capacity for rigorous thought cannot produce an educational system that demands rigorous thought. The system reflects the society, and the society has decided, gradually and unconsciously, that rigor is elitist, that standards are exclusionary, and that self-esteem matters more than competence.

The political dimension makes reform extraordinarily difficult. Education in America is controlled at the local level by school boards that are elected by communities with wildly different values and priorities. Federal policy can set broad mandates but can't control what happens in 13,000 independent school districts. Teachers' unions protect their members but sometimes resist reforms that would improve outcomes. Textbook publishers sell to the largest markets, Texas and California, and the rest of the country gets whatever those two states approve. What results is an education system that satisfies nobody, improves slowly if at all, and produces graduates who are measurably less competent than their counterparts in countries that spend half as much.

• • •

The optimistic case has two parts, and they operate on different timelines.

Short term: educational reform is a choice, not a destiny. Other countries have transformed their educational systems within a generation. Singapore went from a developing-world education system to the top of every international ranking in roughly thirty years through a deliberate national strategy that prioritized teacher quality, rigorous standards, and STEM education. Finland did something similar with a different model. Both demonstrate that declining educational quality is not inevitable. It's a policy failure, and policy failures can be corrected by better policy.

The US could do what Singapore did. Pay teachers like the professionals they should be. Set rigorous national standards and enforce them. Rebuild vocational education with the seriousness Germany gives it. Stop treating STEM education as optional and start treating it as a national security priority, because that's what it is. None of this is mysterious. None of it requires new technology. It requires political will, which, as *Turn Off the TV, Get Off Your Ass, and Do Something* argued, requires an engaged citizenry that demands better.

Long term: cognitive enhancement technology, explored in *The Birth of the Augmented Human* and in Chapters 23 through 25 of this book, may eventually bypass traditional education entirely. If you can install mathematical capability through a neural interface or accelerate learning through pharmacological cognitive enhancement, the entire K-12 through university pipeline becomes obsolete. The question isn't whether that future is coming. It's whether it arrives in time to matter.

In the critical window, the next 15 to 20 years when the other crises in this book need to be addressed, you need educated humans building the solutions. Enhancement might eventually solve the education problem. But the education problem might prevent you from building the enhancement technology. That's the race, and right now, the United States is losing it.

The most dangerous deficit in America isn't financial. It's human capital. Every tool described in the coming chapters, nuclear energy, artificial intelligence, longevity science, genetic engineering, human enhancement, requires people capable of wielding it. The system that produces those people is failing. That failure is solvable, but only if it's treated as what it is: a national security emergency of the first order.

Now for the tools themselves. Starting with the one that solves three crises at once.

Chapter 22: The Nuclear Answer

In 1958, the United States launched the USS Nautilus under the Arctic ice cap and surfaced at the North Pole. The ship ran on a nuclear reactor the size of a truck that could power a city. It didn't need to refuel for years. It produced no exhaust, consumed no oxygen, and could stay underwater indefinitely. The technology was so powerful that the Navy built an entire fleet around it, and the nuclear submarine became the most survivable weapons platform on the planet.

In 1957, the Shippingport Atomic Power Station in Pennsylvania became the first full-scale commercial nuclear power plant in the United States. It worked. It was safe. It demonstrated that nuclear fission could generate electricity for a civilian grid at industrial scale. The technology was American. The engineering was American. The vision was American. For a brief period, the United States led the world in nuclear energy and seemed poised to build an electrical system that would make fossil fuels obsolete.

Then something happened that was neither rational nor inevitable. The country got scared. And for fifty years, it refused to build the one technology that solves three of the biggest crises at the same time.

• • •

The fear came from three accidents across seven decades of nuclear power operation.

Three Mile Island, 1979. A partial meltdown at a Pennsylvania reactor caused by a combination of equipment failure and operator error. The containment building did its job. No significant radiation was released. Nobody died. Nobody got sick. The reactor was a mess, but the safety systems worked exactly as designed: they contained the failure. The actual public health impact was effectively zero.

Chernobyl, 1986. A Soviet-designed reactor with no containment building, operated by a crew conducting an unauthorized safety test with the safety systems deliberately disabled, exploded and burned for days, spreading radioactive contamination across Europe. Roughly 30 people died immediately from radiation exposure. Long-term cancer deaths are estimated in the low thousands, though the exact number is debated. Chernobyl was a catastrophe, but it was a catastrophe produced by a reactor design that no Western country ever used, operated under conditions that would be criminal in any regulatory framework outside the Soviet Union.

Fukushima, 2011. A Japanese reactor complex was hit by a magnitude 9.0 earthquake followed by a 45-foot tsunami. The earthquake shut down the reactors automatically, as designed. The tsunami destroyed the backup power systems that cooled the reactors after shutdown, which was a design flaw that had been identified years earlier and not fixed. Three reactors melted down. Hydrogen explosions breached containment in multiple buildings. Radioactive contamination required the evacuation of roughly 150,000 people. The direct death toll from radiation was one confirmed death, years after the event. The tsunami that caused the meltdown killed roughly 20,000 people, a number that gets lost in the nuclear narrative.

Three accidents. Seventy years. One was contained with zero casualties. One was caused by a design no Western country uses. One was caused by a natural disaster that killed twenty thousand people independently of the reactor. Total confirmed direct fatalities from nuclear power plant accidents in the entire history of the technology: fewer than a hundred.

Coal power kills roughly 800,000 people a year through air pollution. Natural gas kills tens of thousands. Even solar and wind kill more people per kilowatt-hour than nuclear, through manufacturing accidents, installation falls, and the mining of materials. Nuclear energy is, by every statistical measure, the safest form of energy generation humanity has ever developed.

And we stopped building it because we were afraid.

• • •

The case for nuclear power in 2026 isn't just about electricity, though the electricity case is overwhelming. It's about solving three crises that the first four parts identified as existential.

The energy crisis. Solar and wind are real technologies that produce real power. This book is not anti-renewable. But renewables have a fundamental limitation that no amount of investment can overcome: they're intermittent. The sun doesn't always shine. The wind doesn't always blow. When they're not producing, something else has to fill the gap. Currently, that something else is natural gas, which means that every solar panel and wind turbine installed requires a natural gas backup plant sitting idle most of the time, ready to fire up when the weather changes. The "clean" grid is a hybrid grid that still depends on fossil fuels for reliability.

Battery storage is the proposed solution, and it's improving rapidly. But storing enough electricity to power an industrial economy through a week of cloudy, windless weather requires batteries on a scale that doesn't exist and that depends on minerals, lithium, cobalt, nickel, that are concentrated in a handful of countries and processed primarily by China. Chapter 18 described the energy transition as trading one dependency for another. Battery-dependent renewables trade oil dependency for mineral dependency. Nuclear trades both for uranium, which is abundant, widely distributed, and doesn't require Chinese processing.

A nuclear plant runs 24 hours a day, 7 days a week, 365 days a year, regardless of weather, season, or time of day. It provides the reliable baseload power that an industrial economy requires. It doesn't need batteries. It doesn't need backup. It doesn't need Chinese minerals. It just needs fuel rods that last for years and a workforce trained to operate it safely.

• • •

The food crisis. Chapter 14 described the oil-fertilizer-food chain: the Haber-Bosch process runs on natural gas, which provides both the energy and the hydrogen feedstock for ammonia synthesis. As natural gas depletes and prices rise, fertilizer costs follow, food costs follow, and billions of people face hunger.

Nuclear breaks that chain.

Nuclear-powered electrolysis splits water into hydrogen and oxygen. The hydrogen goes into the Haber-Bosch process the same way methane-derived hydrogen does. The ammonia comes out the same. The fertilizer is the same. The food is the same. The only thing that's different is the energy source: a nuclear reactor instead of a gas well.

The technology isn't theoretical. It's been demonstrated. High-temperature reactors can produce hydrogen more efficiently through thermochemical processes that use nuclear heat directly rather than converting it to electricity first. The engineering exists. What doesn't exist is the political will to build the reactors, because the reactors are nuclear and nuclear is scary.

Chapter 14 asked what happens when the chemical that feeds half the world starts running out. The answer is: nothing, if you replace the energy source before it runs out. Nuclear hydrogen production is that replacement. It's sitting on the shelf, fully developed, waiting for someone to decide that feeding 4 billion people is more important than being afraid of a technology that kills fewer people than rooftop solar.

. . .

The water crisis. Chapter 15 described freshwater scarcity as an accelerating geopolitical force, from the Nile Basin confrontation to Iran's collapsing water table to the drying Mediterranean. Desalination, the process of removing salt from seawater to produce fresh water, is the engineering solution. And desalination at scale requires energy. A lot of energy.

Chapter 15 described Israel's desalination achievement: a desert country that turned itself into a water exporter through five major plants along the Mediterranean coast. The technology works. What it requires is energy, and lots of it.

Scaling it to solve the water crises this book described requires energy that renewables can't reliably provide. A desalination plant that shuts down when the wind stops blowing doesn't solve a water crisis. It creates intermittent water supply, which for a city of millions is barely better than no water supply at all. Nuclear-powered desalination provides continuous, reliable freshwater production regardless of weather or season. Build a reactor on the coast, pipe the water inland, and the desert blooms. Not metaphorically. Literally.

Saudi Arabia, the UAE, and the entire Persian Gulf region could transform their water futures with nuclear desalination. So could North Africa. So could the Mediterranean coast. So could any coastal region facing freshwater scarcity. The constraint isn't engineering. It's the fifty-year fear that has prevented the technology from being deployed at the scale the world needs.

• • •

The strategic dimension is the one that should convince even people who don't care about the environment or the food supply.

A country that runs on nuclear energy is energy-independent in the deepest possible sense. Uranium is abundant and geographically distributed. Canada, Australia, Kazakhstan, Namibia, Russia, and the United States all have significant deposits. No single country dominates the supply. There is no uranium OPEC. There is no uranium Strait of Hormuz. The fuel is small, energy-dense, and easily stockpiled. A country with a few years' worth of uranium in storage can ride out any supply disruption, any embargo, any geopolitical crisis, without losing power.

A nuclear-powered United States doesn't need Middle Eastern oil for electricity. Doesn't need to protect Hormuz. Doesn't need Chinese solar panels or rare earth minerals for batteries at the same scale. Can produce its own fertilizer feedstock through hydrogen electrolysis. Can desalinate its own water. Can power its own grid regardless of what happens anywhere else on the planet.

That's genuine energy sovereignty. Not the fake version where we produce enough oil but let the global price determine what we pay. The real version where the energy system is physically independent of every supply chain, every chokepoint, and every adversary.

Chapter 10 argued that energy price independence would be the single most consequential strategic decision the US could make. Nuclear is how you get there. Not through oil export controls, which are a patch on a broken system. Through an energy source that doesn't depend on the system at all.

. . .

The next generation of nuclear technology addresses every objection that the anti-nuclear movement has raised over the past fifty years.

Small modular reactors, SMRs, are factory-built units a fraction of the size of conventional plants. They can be transported by truck or rail and assembled on site. The standardized design eliminates the custom engineering that made traditional nuclear plants expensive and slow to build. Factory production means quality control is consistent and costs decline with volume. An SMR can power a small city, a military base, a desalination plant, or a remote industrial facility. NuScale Power, an Oregon company, became the first SMR design ever certified by the NRC in 2023, and received approval for an uprated 77-megawatt version in May 2025. It remains the only SMR with regulatory approval in the United States. But the path has not been smooth. NuScale's first planned deployment, a project with the Utah Associated Municipal Power Systems at Idaho National Laboratory, was cancelled in late 2023 when cost projections rose above $90 per megawatt-hour and participating utilities backed out. The company is now targeting deployment by 2030 through a new commercialization partner. That timeline, if it holds, means the first American SMR will reach customers roughly 20 years after the technology was first proposed to the NRC.

Molten salt reactors operate at atmospheric pressure rather than the high pressures of conventional designs, which eliminates the risk of a pressure-driven containment breach, the scenario that contributed to Fukushima. The fuel is dissolved in salt, which makes meltdown physically impossible because the fuel is already melted. If the reactor overheats, a freeze plug at the bottom melts, the fuel drains into a passively cooled tank, and the reaction stops. No operator action required. No backup power required. Physics does the work.

Thorium-based fuel cycles use a fuel that is three to four times more abundant than uranium, produces waste that is radioactive for hundreds of years rather than thousands, and is much more difficult to weaponize. The US built and operated a thorium reactor in the 1960s at Oak Ridge National Laboratory. It worked. The program was cancelled for political reasons, not technical ones.

The United States led the world in every one of these technologies. It invented the pressurized water reactor. It pioneered the sodium-cooled fast reactor. It built the first thorium reactor. It developed the first small modular designs. And then it stopped. Regulatory paralysis, public fear, political cowardice, and the lobbying power of the fossil fuel industry combined to freeze the American nuclear industry in place while the rest of the world moved forward.

China is building Gen IV reactors. Its HTR-PM, a high-temperature gas-cooled pebble-bed reactor at Shidaowan in Shandong province, entered commercial operation in December 2023---the world's first Generation IV reactor to do so. It proved during safety testing that it can cool itself down from full power without any human intervention, external power, or emergency systems. China plans to scale it to a 600-megawatt version with six reactor modules, and ultimately to build ten units at the same site. Russia is deploying floating nuclear plants, with the Akademik Lomonosov powering a remote Arctic town since 2020. South Korea is exporting reactor designs. India is pursuing a thorium fuel cycle. The United States, which invented the technology, is fighting about whether to approve a design that other countries are already constructing.

• • •

The obstacle is not physics. It is not engineering. It is not economics, though the current regulatory environment makes nuclear artificially expensive by imposing approval timelines that stretch projects to a decade or more. The obstacle is politics.

And politics can change. The same country that put a man on the moon in eight years, that built the interstate highway system in a decade, that developed the atomic bomb in three years of wartime urgency, can build 200 nuclear reactors in a generation if it decides to. The capability exists. What's missing is the decision.

Nuclear energy is the single technology that at the same time solves the energy crisis, the food crisis, and the water crisis. It provides genuine strategic independence from every chokepoint and every adversary. It is the safest form of energy generation ever developed by any statistical measure. The next-generation designs are smaller, safer, cheaper, and more proliferation-resistant than anything that came before.

The country that led the world in this technology abandoned it out of fear. The countries that are building it now will have energy sovereignty, food security, and water independence in 20 years. The United States can still catch up. But the window is closing, and every year of delay is a year of strategic advantage handed to competitors who weren't afraid to build the future.

• • •

So how does this happen? Not the technology. The deployment. Because the technology already exists, and it has existed for decades. What doesn't exist is a pathway from approved design to operating reactor that takes less than a generation.

238

The Nuclear Regulatory Commission was created in 1975, split off from the Atomic Energy Commission that had both promoted and regulated nuclear power. The AEC had a conflict of interest and the NRC was supposed to fix it by being purely a safety regulator. It succeeded. It is very good at safety. It is also very good at ensuring that nothing gets built. The NRC's licensing process for a new reactor design takes years of review at a cost of hundreds of millions of dollars before a single shovel hits dirt. NuScale spent over a decade and the Department of Energy spent more than $575 million supporting its design through the NRC process. The review of the uprated 77-megawatt design took 22 months, which the NRC called ahead of schedule. For context, China built the entire HTR-PM demonstration plant in roughly the same amount of time it takes the NRC to review the paperwork for one.

The reform agenda is clear even if the politics are not. First, shift the NRC from a prescriptive regulatory model to a performance-based one. Instead of specifying exactly how a reactor must be designed, down to the thickness of individual pipes, define the safety outcomes the reactor must achieve and let the designers figure out how to meet them. This is how aviation works. The FAA certifies aircraft based on performance standards, not by dictating wing shapes. Second, create a standardized licensing pathway for factory-built SMRs so that once a design is approved, individual deployments don't each require a separate multi-year review. The whole point of a modular reactor is that it's the same every time. Reviewing the same design fifty times defeats the purpose. Third, reform the emergency planning zone requirements. Current NRC rules require a 10-mile emergency planning zone around every reactor, which makes siting near population centers or industrial facilities almost impossible. The new SMR designs, which are physically incapable of the kind of meltdown that created those rules, should have site-boundary emergency planning zones scaled to their actual risk profile. NuScale's design already qualifies for this, but the regulatory framework hasn't caught up.

The financing problem is equally real. Nuclear plants are expensive to build and cheap to operate. The capital cost is front-loaded. A utility that wants to build a nuclear plant has to spend billions before generating a single kilowatt-hour of revenue, and the construction timeline means the money is tied up for years. This is why the UAMPS project collapsed: the projected cost per megawatt-hour kept rising as the timeline stretched, and the participating utilities backed out. The fix is a federal loan guarantee program specifically for nuclear construction, modeled on the Rural Electrification Administration that brought power to rural America in the 1930s and 1940s. The government doesn't build the reactors. It guarantees the loans, which drops the interest rate, which drops the total cost, which makes the projects viable. The Department of Energy's Loan Programs Office already has authority to do this. It needs the political mandate to do it at scale: not one reactor at a time, but a national program of 200 reactors over 20 years, with standardized designs, streamlined permitting, and a commitment that doesn't evaporate with the next election cycle.

Nuclear is the foundation. Without it, nothing else in Part Six works at scale. But a foundation needs something built on top of it, and the technology that turns nuclear energy from a power source into a civilizational advantage is the one that makes everything faster, smarter, and more efficient than any human institution can manage on its own.

Chapter 23: The AI Multiplier

Watch what Ukraine is doing with military AI. It is genuinely exciting — a smaller country with fewer resources is innovating faster in autonomous systems and battlefield AI than anyone expected. Drone swarms coordinated by AI. Real-time targeting and decision loops that operate faster than human reaction. Ukraine is running the real-world laboratory that the US military has been theorizing about for years. The scary part is that the US military is far behind where it should be given our resources. We have the money and the talent. We do not have the urgency. Ukraine has urgency because losing is not an abstraction. That gap needs to close.

In November 2023, a pharmaceutical researcher named Alex Zhavoronkov watched an AI system at his company, Insilico Medicine, identify a drug candidate for idiopathic pulmonary fibrosis in eighteen months. The traditional process takes four to six years and costs hundreds of millions of dollars. Insilico's AI analyzed the disease's molecular structure, generated novel compounds, predicted which ones would bind to the target protein, filtered for toxicity, and produced a candidate that entered Phase II clinical trials. The drug may or may not work. What matters is the timeline. Eighteen months from target identification to human trials, at a fraction of the traditional cost. One disease. One company. One AI system that did in a year and a half what used to take a decade.

Now multiply that across every problem in this book.

A force multiplier is a military term for something that makes an existing force more effective without increasing its size. Night-vision goggles are a force multiplier: the same soldiers, but now they can fight in the dark. Satellite communications are a force multiplier: the same units, but now they can coordinate in real time. Precision-guided munitions are a force multiplier: the same aircraft, but now each bomb hits its target instead of cratering an empty field.

Artificial intelligence is the largest force multiplier in human history. Not just for the military, though the military applications are staggering. For everything. Every crisis in this book. Every tool described in Part Six. Every institutional capacity that the PE chapter described being hollowed out. AI doesn't replace the things that are broken. It makes the things that still work dramatically more effective, and it makes rebuilding the broken things dramatically faster.

This chapter is not about artificial general intelligence. It's not about the singularity. It's not about whether AI will become sentient or replace all human jobs or enslave humanity. Those are interesting questions for a different book. This chapter is about what AI can do right now, and in the next ten years, for the specific problems this book has identified.

• • •

Start with the demographic crisis from Chapter 12, because it's the one where AI's impact is most immediate and most measurable.

Japan has been testing the proposition that automation can compensate for a shrinking workforce for two decades. The results are mixed but directionally positive. Japanese factories are among the most automated on the planet. Robots do work that humans used to do, and the output per remaining worker has increased. Japan's GDP per capita hasn't collapsed despite a shrinking population because productivity gains have partially offset the decline in labor.

AI takes that proposition and accelerates it by an order of magnitude. Factory robots replaced human muscles. AI replaces human cognitive work: analysis, decision-making, planning, pattern recognition, optimization. A company that used to need fifty analysts to process supply chain data needs five analysts and an AI system that does the work of the other forty-five, faster, more accurately, and without taking holidays.

The math matters here. If AI can increase the productivity of the average knowledge worker by 30 to 50 percent---McKinsey estimates generative AI could automate 60 to 70 percent of current work activities, and Goldman Sachs found median productivity gains of around 30 percent in specific deployments like software development and customer service---then a country that loses 20 percent of its workforce to demographic decline doesn't lose capacity. The remaining workers, augmented by AI, produce the same output or more. The demographic cliff becomes a demographic slope. Still a problem, but a manageable one rather than a civilizational crisis.

This doesn't work for every type of labor. AI can't weld a submarine hull. It can't change a hospital patient's bedpan. It can't fix a pothole or install a heat pump. The physical work that Chapter 21 described being undersupplied still requires human hands, and no amount of AI compensates for the vocational training pipeline that was dismantled. But for the enormous share of economic activity that consists of information processing, analysis, communication, and decision support, AI is a genuine substitute for human labor at a ratio that changes the demographic equation.

But the same math that solves the demographic problem creates a political one. If AI makes each remaining worker 30 to 50 percent more productive, that means 30 to 50 percent fewer workers are needed to produce the same output. In a country with a shrinking workforce, those numbers balance: fewer workers, each doing more, total output stable. In a country where the workforce is not yet shrinking, or in sectors where the decline hasn't hit, the math produces unemployment. Not the gentle, gradual kind that retraining programs can absorb. The fast, concentrated kind that hits specific professions, entire industries, and particular regions all at once.

Customer service, data analysis, legal research, medical coding, financial modeling, content writing, software testing, truck driving, warehouse logistics. The white-collar and service-sector jobs that replaced the manufacturing jobs that went to China are now being replaced by algorithms that work faster, cheaper, and without health insurance. A country that is already politically fractured along lines of economic anxiety does not become less fractured when millions of knowledge workers discover that their college degrees no longer protect them.

The populism that Chapter 33 describes is fed by exactly this kind of displacement. AI is the multiplier. It multiplies the solutions. It also multiplies the anger.

• • •

The military applications are the ones that connect most directly to Part One's analysis of the Iran war.

The cost-exchange ratio problem, $12 million interceptors versus $300,000 missiles, is at its core a problem of expensive, handcrafted weapons trying to defend against cheap, mass-produced threats. AI changes that equation in two ways.

First, autonomous weapons systems can be cheap and numerous. A drone swarm of a thousand AI-controlled units, each costing a few thousand dollars, each capable of identifying and engaging targets independently, inverts the cost-exchange ratio. The defender sends cheap things to kill expensive things. The attacker has to spend millions defending against thousands of expendable threats that each carry just enough lethality to do damage.

The Pentagon's Replicator program, launched in August 2023 with $1 billion in funding, was explicitly designed to field autonomous systems in the thousands within two years. By its August 2025 deadline, the program had delivered hundreds of drones to warfighters, not the multiple thousands promised. AeroVironment's Switchblade-600 loitering munitions and Anduril's Ghost-X drones were among the systems fielded.

The concept is right. The execution demonstrates the same acquisition dysfunction that keeps recurring: the Pentagon announced an urgent program, classified the details, underfunded the implementation, and missed the target. A second phase focused on counter-drone systems is underway, and the Trump administration launched a separate Drone Dominance Program aimed at producing 300,000 small attack drones. The aspiration to turn American industrial capacity into cheap mass is real. Whether the acquisition system can deliver it remains the question.

Second, AI-powered sensor fusion and decision support make existing weapons more effective. An Aegis destroyer that currently requires human operators to evaluate threats, assign weapons, and manage the engagement sequence could have AI handling the entire process in milliseconds. The human stays in the loop for the authorization to fire, but the AI does the sensing, tracking, threat evaluation, and weapons assignment at machine speed. The same ship, the same number of VLS cells, but each interceptor is more likely to hit its target and less likely to be wasted on a decoy. The magazine lasts longer because fewer shots miss.

Directed energy weapons, the lasers described in Chapters 1 and 2, need AI targeting systems to be effective against swarm attacks. A human operator can track and engage one or two threats at a time. An AI can manage a laser turret engaging dozens of targets per second, prioritizing by threat level, predicting trajectories, and allocating power across multiple emitters. The technology that makes cheap missile defense possible, the laser, requires the technology that makes fast targeting possible, the AI, to work at the speed the battlefield demands.

. . .

The food crisis. Chapter 14 described precision agriculture as one of the solutions to the fertilizer dependency, and AI is what makes precision agriculture work at scale.

Current farming applies fertilizer the way a medieval physician applied leeches: broadly, uniformly, and wastefully. A farmer spreads nitrogen across a field based on rules of thumb, historical practice, and whatever the farm supply dealer recommends. Most of the nitrogen doesn't reach the crops. It runs off into waterways, creating dead zones in the Gulf of Mexico and algae blooms in Lake Erie. It's a massive inefficiency that wastes expensive fertilizer and damages the environment at the same time.

AI-driven precision agriculture uses satellite imagery, soil sensors, weather data, and machine learning to map the nitrogen needs of a field at the resolution of individual meters. The AI knows which patches of soil are nitrogen-depleted and which are saturated. It directs variable-rate application equipment to put exactly the right amount of fertilizer in exactly the right place. What results is 30 to 50 percent less fertilizer used for the same or better yields. Applied globally, that's a 30 to 50 percent extension of the natural gas supply that feeds the Haber-Bosch process. Decades of additional runway before the depletion curve becomes critical.

The same AI systems optimize planting schedules, irrigation timing, pest management, and harvest logistics.

246

Each optimization is a few percentage points of efficiency. Stack them across the entire agricultural system and you get a transformation in food production that partially decouples it from those resource constraints. Not fully. You still need the fertilizer. But you need much less of it, and the food system becomes much more resilient to supply disruptions.

· · ·

Drug discovery and pandemic response. Chapter 16 described the vulnerability to the next pandemic and the inadequacy of the response infrastructure. AI is the technology that makes the response faster than the pathogen.

AlphaFold, DeepMind's protein structure prediction system, solved in 2020 a problem that structural biologists had been working on for fifty years: predicting the three-dimensional shape of a protein from its amino acid sequence. It won the CASP14 competition by such a margin that the organizers declared the protein folding problem in practice solved. By 2022, DeepMind had released predicted structures for over 200 million proteins, nearly every protein known to science, a dataset that would have taken experimental methods billions of years to produce. Its creators, Demis Hassabis and John Jumper, won the 2024 Nobel Prize in Chemistry. Over three million researchers in 190 countries now use AlphaFold as a standard tool. The implications for drug discovery are enormous. If you know the shape of a virus's surface proteins, you can design molecules that bind to them and neutralize them. AI can screen billions of potential drug compounds against a viral target in days, a process that used to take years of laboratory work.

mRNA vaccine development, which was already fast during COVID, gets faster with AI assistance. The AI can predict which parts of a new pathogen's protein structure will make the best vaccine targets. It can optimize the mRNA sequence for stability and immune response. It can model the immune system's likely reaction and predict efficacy before clinical trials begin. The goal, a vaccine candidate within days of a new pathogen being sequenced, is achievable with current AI capability. The bottleneck shifts from discovery to manufacturing and distribution, which are logistics problems rather than science problems.

AI-powered epidemiological surveillance, monitoring wastewater data, hospital admission patterns, genomic sequences from clinical labs, social media chatter about symptoms, can detect an outbreak before the traditional public health system recognizes it. The early warning that was missed for COVID, the weeks between the first cases in Wuhan and the international response, could be compressed to days with AI systems that don't take weekends off and don't wait for bureaucratic approval to raise an alarm.

• • •

Nuclear deployment. Chapter 22 described the political and regulatory obstacles to nuclear power. AI addresses the practical ones.

Reactor design optimization that used to take engineering teams years of iterative modeling can be done in weeks with AI-assisted computational tools. The design space for advanced reactors, including the small modular reactors and molten salt designs described in the previous chapter, is vast. AI can explore that space thousands of times faster than human engineers, identifying optimal configurations for safety, efficiency, cost, and constructability.

AI-managed construction scheduling and quality control can compress the timeline for building nuclear plants. The regulatory approval process, which currently stretches to a decade or more, involves massive documentation requirements that AI can automate. Not the judgment calls, which should remain human, but the paperwork, the compliance documentation, the safety analysis reports that run to thousands of pages. If AI reduces the documentation timeline from years to months, the overall project timeline compresses proportionally.

AI-operated reactors are safer than human-operated ones. The Fukushima failure chain included delays in human decision-making at critical moments. An AI monitoring system doesn't panic, doesn't freeze, doesn't defer to authority when it detects an anomaly. It responds at machine speed with optimal actions. Every reactor safety system described in Chapter 22 works better with AI supervision.

. . .

The education crisis. Chapter 21 described the American educational system producing fewer competent graduates every year. AI provides a partial bypass, and the word "partial" matters.

AI tutoring systems can provide personalized instruction to every student at every level, adapting in real time to the student's comprehension, pace, and learning style. A human teacher managing thirty students can't do this. An AI system can provide the equivalent of one-on-one tutoring to every student at the same time. Studies of early AI tutoring deployments show significant improvement in learning outcomes, particularly for students who are behind grade level. The students who most need help are the ones who benefit most from AI assistance.

This doesn't replace teachers. It replaces the worst version of teaching: the lecture to a room of thirty students at thirty different comprehension levels, where the teacher pitches to the middle and hopes the top and bottom thirds figure it out on their own. AI handles the differentiation that no human teacher has time for, freeing the teacher to do the things humans are better at: mentoring, motivating, explaining the why behind the what, and identifying students who need emotional support rather than just academic instruction.

The bypass is partial because AI can't fix a culture that doesn't value learning. It can't make a student care about mathematics. It can't replace a home environment that supports education. It can't solve the funding, governance, and political dysfunction that Chapter 21 described. What it can do is make the instruction that does happen dramatically more effective, which buys time for the structural reforms that take a generation to implement.

. . .

The geopolitical competition for AI leadership is, along with the nuclear race, the defining technology competition of the century. And the United States currently holds the lead, but the lead is narrower than most Americans realize and more fragile than the technology sector acknowledges.

American advantages: the best AI research labs in the world, including Google DeepMind, OpenAI, Anthropic, and Meta's AI research division. The best universities producing AI talent. The deepest venture capital ecosystem funding AI startups. The most advanced chip designs from Nvidia and AMD. A culture of open research and rapid iteration that produces breakthroughs at a pace authoritarian systems can't match.

American vulnerabilities: the chips are designed in America but mostly manufactured in Taiwan, at TSMC. If the Taiwan Strait becomes a combat zone, the AI chip supply chain breaks. The training data for AI models is increasingly subject to legal and regulatory constraints that don't apply in China. The energy required to train and run large AI models is enormous and growing, which connects directly to the nuclear question. And the talent pipeline depends heavily on foreign researchers who face immigration barriers.

China's advantages: state-directed investment at a scale no private company can match. A huge domestic market for AI deployment. Fewer regulatory constraints on data collection and use. A government that treats AI development as a strategic national priority rather than leaving it to market forces. And a willingness to deploy AI systems at scale in ways that Western societies find ethically uncomfortable but that produce rapid improvement through real-world feedback.

The competition is real, it's accelerating, and the stakes are as high as any technology race in history. The country that leads in AI leads in military capability, economic productivity, scientific discovery, and the ability to manage every other crisis described here. Losing the AI race isn't like losing a trade competition. It's like losing the nuclear race in 1945: a disadvantage that reshapes the entire strategic landscape for generations.

But deploying AI at scale requires confronting two problems that the breathless coverage of chatbots and image generators tends to ignore: energy and governance.

The energy problem is staggering. Training a single frontier AI model consumes as much electricity as a small city uses in a year. Running those models at scale, answering millions of queries per day, processing satellite imagery, managing autonomous weapons systems, optimizing power grids, requires data centers that draw hundreds of megawatts of continuous power. The International Energy Agency projects that global data center electricity consumption could double by 2026. Every major AI company is scrambling for power: Microsoft signed a deal to restart the Three Mile Island reactor. Google and Amazon are investing in small modular reactors. Meta is trying to build a nuclear-powered data center. The connection between this chapter and the previous one is not metaphorical. AI runs on electricity, and the scale of AI deployment being planned requires a massive expansion of electrical generation that renewables alone cannot provide.

Nuclear power and AI are not separate stories. They are the same story told from different angles.

The governance problem is more diffuse and more dangerous. AI systems that can design pathogens, crack encryption, generate convincing disinformation at scale, or autonomously target and kill human beings require some framework of rules. The question is what kind. The European Union has chosen prescriptive regulation, passing the AI Act in 2024 with detailed requirements for transparency, risk assessment, and prohibited uses. China has chosen state control, requiring AI companies to register algorithms and submit training data for government review. America has chosen almost nothing. Executive orders have come and gone with changing administrations. Congress has held hearings and produced no legislation. The AI companies are regulating themselves, which means they are not being regulated at all. What results is that the most powerful technology since nuclear weapons is being deployed globally with less governmental oversight than a new pharmaceutical drug or a children's toy.

The argument against regulation is that it will slow American companies and hand the advantage to China. The argument for it is that ungoverned AI will eventually produce a catastrophe, whether a bioweapon designed by an AI system, a flash crash in financial markets triggered by competing algorithms, or an autonomous weapons system that kills the wrong people, that makes the political backlash against AI far worse than any regulation would have been. The smart play is targeted regulation that addresses the actual risks without strangling the industry: mandatory safety testing for frontier models, export controls on the most capable chips, liability rules for autonomous systems that kill people, and a national AI safety institute with the authority and funding to keep up with the technology. None of this is impossible. All of it requires Congress to act, which is the bottleneck that keeps appearing in every chapter.

Everything above describes what AI does for us. Now for what it does against us, because the other side has the multiplier too.

Chapter 11 described foreign information manipulation campaigns that flood democratic information environments with contradictory narratives until the cost of determining what is real exceeds what most citizens will pay. AI makes that weapon orders of magnitude more powerful. A state-sponsored troll farm employing hundreds of people can now be replaced by a single AI system generating thousands of posts per hour, each tailored to a specific audience, written in fluent local idiom, indistinguishable from authentic human speech. Deepfake videos of political leaders saying things they never said can be produced in minutes. Synthetic news articles from fabricated outlets can be generated and seeded across platforms at industrial scale. The information environment that was already polluted becomes toxic. The cost-exchange ratio from Chapter 2 applies: generating disinformation with AI costs almost nothing, detecting and debunking it costs everything. The defender is always behind because the attacker can produce faster than the defender can verify.

China's AI-powered domestic surveillance state is the authoritarian version of the multiplier. Facial recognition, social media analysis, behavioral prediction, the social credit system. It makes repression more efficient the same way AI makes American drug discovery more efficient. The same technology, applied to different purposes by different systems. And China is not keeping the surveillance AI for domestic use. It is exporting it to governments across the developing world through the Digital Silk Road described in Chapter 18, giving authoritarian regimes the tools to monitor their populations with a sophistication that the Stasi could not have imagined. The AI multiplier multiplies whatever you point it at. Point it at freedom and you get more freedom. Point it at control and you get more control. The competition between the American and Chinese AI models is, at bottom, a competition over what the multiplier gets aimed at.

The concentration problem is the one this book's readers should recognize immediately, because it is the PE thesis wearing a hoodie. As of 2026, frontier AI development is controlled by fewer than half a dozen companies. The compute required to train a frontier model costs hundreds of millions of dollars. The chips come overwhelmingly from one designer, Nvidia. The fabrication happens at one foundry, TSMC. The cloud infrastructure runs on three platforms. This is not an open market. It is an oligopoly with the same structural characteristics that Chapter 20 described in defense contractors and hospital chains: a small number of players controlling a critical resource, with barriers to entry so high that competition is theoretical rather than real.

If AI is the multiplier that makes everything in Part Six work, then the companies that control AI are the gatekeepers of the entire stack. The history of concentrated control over critical infrastructure, from Standard Oil to the defense primes, suggests that the gatekeepers will eventually optimize for their own interests rather than the public's. That is not a prediction about the intentions of the people running these companies. It is a prediction about incentive structures, and the incentive structures are identical to the ones that produced the extraction patterns this book has documented since Chapter 20.

• • •

AI is not a solution to any single problem described here. It's a multiplier that makes every solution more effective and every problem more manageable. But the multiplier is neutral. It amplifies whatever it touches. Applied to the solutions, it accelerates them. Applied to the problems, it accelerates those too. Applied to the information environment, it degrades it. Applied to surveillance, it perfects it. Applied to a concentrated industry, it concentrates it further. The question is not whether AI is good or bad. The question is who controls it, what they aim it at, and whether the democratic systems that are supposed to govern it can keep up with the pace of deployment. So far, they cannot. Nuclear deployment goes faster with AI. Food production gets more efficient. Drug discovery accelerates. Military systems become more capable. Educational outcomes improve. Manufacturing gets more productive. The demographic cliff becomes a slope. The workforce shortage becomes less acute.

The multiplier only works if you have something to multiply. AI applied to a hollowed-out industrial base multiplies the hollowness. AI applied to an education system that doesn't produce STEM graduates helps the STEM graduates you have but doesn't create new ones. AI applied to a military with empty missile magazines makes the remaining missiles more accurate but doesn't fill the magazines. The foundation has to be rebuilt. AI makes the rebuilding faster. It doesn't substitute for it.

That's the correct framing for AI not a savior, but an accelerant. Pour it on the solutions and they work faster. Pour it on the problems and the problems get diagnosed faster. But someone still has to do the building. Someone still has to make the decisions. Someone still has to show up and weld the hull.

Pause here, because the next five chapters cover technologies that sound like science fiction and aren't. What connects them is something the individual chapters can't show: the technologies compound. Nuclear powers the AI. The AI accelerates the drug discovery. The drugs extend productive lifespans. The extended lifespans ease the demographic crisis. The eased demographics buy time for the genetic engineering to create disease-resistant populations. Each tool makes the others more powerful, and each missing tool weakens the rest. That compounding is why Part Six exists as a unit rather than a grab bag of interesting technologies. The question is not whether any single tool works. They all work. The question is whether the country deploys them as a system, the way China is doing, or as disconnected programs that compete for scraps.

And there are not enough people to do it. Chapter 12 made that case. The workforce is shrinking. The experienced workers are retiring. The demographic cliff is real. AI compensates for some of the loss. But there is another approach that addresses the problem from the opposite direction: instead of replacing the workers who leave, keep them productive longer. Much longer.

Chapter 24: The Long Life

The demographic crisis operates on a simple assumption: that human productive lifespan is roughly fixed. You're born, you're dependent for about 20 years, you're productive for about 40 years, you decline and become dependent again for the last 15 to 20, and you die. The dependency ratio, the number of non-working people each worker supports, is the formula that breaks when fertility falls. Fewer workers. Same number of retirees. The math doesn't work.

But the math is built on biology. And biology is about to change.

. . .

Aging is not a mystery anymore. That's the sentence that separates the 21st century from every previous one. For all of human history, aging was treated as an inevitability: the body wears out, the systems fail, you die. Religion offered explanations. Philosophy offered acceptance. Medicine offered palliatives. Nobody offered a mechanism because nobody understood the mechanism.

Now they do. Not completely, not in every detail, but the major pathways of biological aging have been identified, characterized, and in some cases experimentally reversed in animal models. Aging is a set of biological processes, not a mystical fate. And biological processes can be modified.

The nine hallmarks of aging, identified in a landmark 2013 paper and expanded since, include genomic instability, telomere attrition, epigenetic alterations, loss of proteostasis, deregulated nutrient sensing, mitochondrial dysfunction, cellular senescence, stem cell exhaustion, and altered intercellular communication. Each of these is a specific biological process with specific molecular mechanisms that can be targeted with specific interventions.

If that sentence sounded like jargon, here's the translation: scientists have identified the nine things that cause your body to deteriorate, and they're working on ways to slow, stop, or reverse each one. Some of the interventions are already in clinical trials. Some are in animal testing. Some are still in the lab. None of them are science fiction.

• • •

The interventions that are closest to deployment are the least dramatic and the most immediately useful.

Senolytics are drugs that selectively kill senescent cells, the zombie cells that accumulate with age, stop dividing, but refuse to die. They sit in your tissues, secreting inflammatory molecules that damage neighboring healthy cells and contribute to every age-related disease from arthritis to Alzheimer's. Removing them in animal models has produced dramatic results: improved tissue function, reduced inflammation, and extended healthy lifespan. Multiple senolytic drugs are in human clinical trials as of 2026. The combination of dasatinib, a leukemia drug, and quercetin, a plant compound found in onions and apples, was the first senolytic cocktail tested in humans and has shown promise in early trials for diabetic kidney disease and idiopathic pulmonary fibrosis. Fisetin, a flavonoid found in strawberries, is being tested in trials at the Mayo Clinic for its ability to clear senescent cells in older adults. Unity Biotechnology, a biotech company backed by Jeff Bezos and Peter Thiel among others, is developing senolytic eye drops for age-related macular degeneration and injectable senolytics for joint disease. None of these has been approved yet. The field is young. But the pipeline is real, the mechanism is understood, and the first approved senolytic treatments are likely within five to ten years.

The strategic implications of this science are not abstract. If senolytic treatments extend healthy lifespan by even five to ten years, the economic impact dwarfs any technology investment described in this book. Each year of extended healthy productivity for the average American worker represents tens of thousands of dollars in economic output, tax revenue, and reduced healthcare costs. Multiply that by millions of workers and the numbers become staggering. The demographic cliff from Chapter 12 flattens. The Social Security trust fund depletion timeline extends. The skilled workforce that Chapter 3 said was retiring too fast stays on the job longer. A 70-year-old nuclear welder with the physical capacity of a 55-year-old is worth more to the shipyard than any recruitment bonus for a 22-year-old who needs seven years of training before becoming productive.

Rapamycin analogs target a pathway called mTOR that regulates cell growth and aging. Rapamycin itself, an immunosuppressive drug used in organ transplant patients, has been shown to extend lifespan in every animal model tested, from yeast to mice. The challenge is the side effects of immunosuppression, which are unacceptable for healthy people taking the drug for longevity. Analogs that provide the aging benefits without the immune suppression are in development.

Metformin, a diabetes drug that has been in use for decades and costs pennies per pill, has shown associations with reduced all-cause mortality in diabetic patients compared to non-diabetic controls. Read that again: diabetics taking metformin lived longer than healthy people not taking it. The TAME trial, Targeting Aging with Metformin, is testing whether metformin extends healthspan in non-diabetic adults. If it works, the first clinically proven anti-aging drug could be a generic that costs almost nothing.

Epigenetic reprogramming, based on the Yamanaka factors discovered in 2006, has demonstrated in laboratory settings that aged cells can be partially reprogrammed to a younger state without losing their specialized identity. The cells don't become stem cells. They stay what they are, a skin cell stays a skin cell, a muscle cell stays a muscle cell, but they function like younger versions of themselves. Altos Labs, launched in 2022 with $3 billion in funding backed by Jeff Bezos among others, is pursuing this approach with a roster of researchers that includes four Nobel laureates. Shinya Yamanaka himself, who discovered the reprogramming factors, sits on the scientific advisory board. So are labs around the world. The timeline for clinical application is uncertain, but the principle has been demonstrated.

• • •

Here is where the demographic math changes.

The dependency ratio assumes that a 70-year-old is a dependent. Consuming resources. Drawing a pension. Requiring healthcare. Not producing economic output. The current Social Security system assumes you work until 65 or 67 and then stop. The entire pension and retirement infrastructure of every developed nation is built on the assumption that people become unproductive at a fixed age and then decline until they die.

If longevity science extends the healthy, productive years, the math inverts.

A 70-year-old who functions physically and cognitively like a 50-year-old is not a dependent. That person is a worker, a taxpayer, a consumer, an experienced professional with decades of institutional knowledge. That person is an asset, not a liability. If the interventions described above can add 15 to 20 years of healthy productivity to the average lifespan, the dependency ratio doesn't just stabilize. It improves. Each person works longer, pays taxes longer, consumes healthcare less intensively during the extended healthy period, and compresses the period of decline into a shorter window at the end.

The economic impact is staggering. A person who works productively from 20 to 85 instead of 20 to 65 generates 30 percent more lifetime economic output, pays 30 percent more lifetime taxes, and draws pension benefits for a potentially shorter period relative to their productive years. Scale that across an entire population and the demographic crisis doesn't just moderate. For the countries that adopt longevity technology aggressively, it reverses.

. . .

The strategic implications are at least as significant as the economic ones.

Chapter 3 described the workforce crisis in the shipyards: nuclear-qualified welders take five to seven years to train, and when they retire the knowledge leaves with them. If those welders can work productively for 15 additional years, the workforce shortage eases without training a single new person. The tribal knowledge that walks out the door when a 65-year-old retires stays in the institution for another decade and a half, available to train the next generation.

The military faces the same dynamic. Experienced officers, pilots, submarine commanders, intelligence analysts, all of them represent decades of training and operational experience that can't be replaced quickly. A fighter pilot who reaches the physical limits of service at 45 or 50 represents a massive sunk investment that is discarded because the body can't keep up with the aircraft. If longevity interventions maintain the physical capability, the pilot flies longer, the investment pays off longer, and the force retains experience that takes decades to develop.

The research pipeline benefits even more directly. The scientists and engineers who develop the technologies described in this book, nuclear reactors, AI systems, genetic engineering tools, enhancement devices, require years of education and years of productive research before they make their breakthroughs. Einstein published special relativity at 26, but he was an outlier. Most significant scientific contributions come from researchers in their 30s through 50s who have accumulated enough knowledge and experience to see the connections others miss. Extend that productive window by 15 to 20 years and each researcher produces more breakthroughs, trains more students, and advances their field further.

. . .

The counterarguments are real and deserve honest engagement rather than dismissal.

Retirement systems. If people live and work longer, the retirement age has to adjust. Social Security was designed when the average man died at 63; the retirement age of 65 was set above life expectancy, not below it. If healthy lifespan extends to 90 or 100, retiring at 67 and drawing benefits for 30 years breaks the system just as thoroughly as the current demographic crisis, only from the other direction. The retirement age would need to adjust upward with healthy lifespan extension, which is politically difficult but mathematically necessary.

Inequality. If longevity treatments are expensive and available only to the wealthy, they create a world where rich people live to 100 in good health while poor people die at 75 the way they always have. The longevity divide would be the starkest inequality in human history, making the current gap between rich and poor look trivial by comparison. The ethical imperative is to make these treatments universally affordable, which means the metformin-style cheap interventions matter more than the Altos Labs-style billion-dollar moonshots, at least in the near term.

Resource consumption. More healthy years means more consumption. More food, more energy, more water, more materials. On a planet already straining under 8 billion people, extending the lifespan of those 8 billion puts additional pressure on every resource described in this book. The counter to this counter is that the same period that extends lifespan also sees fertility decline, so the population doesn't grow. It just ages more slowly. The demographic transition produces a smaller, older, longer-lived population rather than a larger, younger one. Whether that's net positive or negative for resource consumption depends on how efficiently the longer-lived population uses resources, which connects back to the AI and nuclear energy arguments.

. . .

The geopolitical dimension is the one that accelerates everything.

If the United States develops and deploys longevity technology aggressively, its workforce stays productive longer, its institutional knowledge persists, its innovation pipeline extends, and its military retains experienced personnel. The demographic advantage it already holds over China, courtesy of immigration, widens further. China's demographic cliff becomes steeper if the US extends productive lifespans while China's aging population continues to contract.

Conversely, if China develops longevity technology first, it could partially offset its catastrophic fertility decline. The 200 million workers that official projections say China loses by 2050 could be kept productive for another decade, buying time for automation and other adjustments. The race to develop longevity interventions is as strategically significant as the AI race, and for the same reason: the winner gets a structural advantage that compounds over decades.

The countries that get old before they get rich, most of the developing world, face the worst version of this dynamic. They can't afford longevity treatments at scale, their populations age without the productivity gains, and the gap between them and the nations that can afford the technology widens into a chasm. Longevity science, like every technology in this book, is both a tool for progress and a mechanism of divergence. Who gets it matters as much as what it does.

The deployment path for longevity science runs through a bottleneck that is both regulatory and conceptual. The FDA does not recognize aging as a disease. It approves drugs for specific diseases: cancer, diabetes, Alzheimer's. It does not approve drugs for "aging." This means every longevity intervention has to be framed as a treatment for something else. Senolytics get tested as treatments for osteoarthritis or idiopathic pulmonary fibrosis, not as treatments for the underlying aging process that causes both. Metformin gets tested through the TAME trial as a way to delay the onset of multiple age-related diseases at the same time, because that framing is the closest the FDA will come to approving a drug for aging itself.

The TAME trial, designed to enroll 3,000 adults aged 65 to 79 across fourteen US sites for six years, has been trying to launch since 2015. As of 2026, it is still only partially funded, held back because metformin is a generic drug that costs pennies per pill and no pharmaceutical company stands to profit from proving it works for aging.

The National Institute on Aging contributed $5 million toward an estimated cost of $45 to $70 million. ARPA-H is now reportedly involved in advancing it. Meanwhile, Eli Lilly is planning a similar trial design using its GLP-1 agonist, because Lilly can profit from that result. The commercial incentives of the pharmaceutical industry and the public health imperative of longevity research are misaligned, and the TAME trial is the most visible casualty.

The fix is the same conceptual shift that nuclear energy requires: treat aging as what it is. Reclassify it. The FDA needs to establish aging as an indication, a condition that drugs can be approved to treat. This would unlock pharmaceutical investment in longevity drugs the way the Orphan Drug Act of 1983 unlocked investment in rare disease treatments: by creating a regulatory pathway that makes the economics work. If aging were an indication, every drug company in the world would be racing to be first to market with an approved longevity treatment, because the market would be every human being over 50. The TAME trial is designed to create exactly this precedent. If it shows that metformin delays the onset of multiple age-related diseases at the same time, the FDA would have the evidence base to recognize aging as a treatable condition. That single regulatory decision would do more to accelerate longevity science than any individual drug discovery.

• • •

Longevity science rewrites the demographic model but doesn't rewrite the genome. It extends the productive lifespan of existing humans with existing capabilities. There is a technology that goes further. One that changes the capabilities themselves, at the most fundamental level possible.

Not extending life. Rewriting the code it runs on.

Chapter 25: The Rewritten Code

Every technology discussed so far in Part Six operates on existing human beings. Nuclear energy powers the civilization they live in. AI multiplies the work they do. Longevity science extends the time they do it. All of them accept the human organism as given and work around its limitations.

Genetic engineering doesn't work around the limitations. It removes them. And unlike every other technology in Part Six, the changes are permanent. Not permanent as in "lasts a long time." Permanent as in "passes to your children, and their children, and every generation after that, forever." In 2018, a Chinese scientist named He Jiankui proved this was not theoretical. He used CRISPR to edit the genomes of twin girls. The international community condemned him. China sentenced him to three years in prison. He served his time, walked out, and opened a new lab. The technology was used. On humans. By a single researcher with modest resources. The barriers to genetic modification of humans are not technological. They are ethical, legal, and regulatory. And those barriers are only as strong as the weakest enforcement regime on the planet.

That makes it the most powerful and the most dangerous tool in this book. And the most inevitable.

. . .

CRISPR-Cas9, the gene editing system that revolutionized biology when it was characterized in 2012, works like a molecular word processor. You give it a guide RNA that matches a specific sequence in the genome. The Cas9 protein finds that sequence, cuts the DNA at that precise location, and the cell's repair machinery either disables the gene or inserts a new sequence you've provided. Cut, paste, replace. At the level of individual nucleotides, the letters of the genetic code.

The technology is cheap. A CRISPR experiment that would have been impossible at any price twenty years ago can be done for a few hundred dollars today. The guide RNAs can be ordered online. The Cas9 protein is commercially available. A competent graduate student with basic molecular biology training can edit a gene in a cell line in a week. The barriers to entry have collapsed, which is both the technology's greatest promise and its greatest danger.

Since 2012, CRISPR and its successor systems have been used to edit genes in plants, animals, bacteria, and humans. Sickle cell disease, caused by a single mutation in the hemoglobin gene, has been effectively cured in clinical trials using CRISPR-based therapies. In December 2023, the FDA approved Casgevy, developed by Vertex Pharmaceuticals and CRISPR Therapeutics, making it the first CRISPR-based gene therapy approved for use in the United States. A disease that has caused suffering for millions of people across thousands of years, eliminated by editing a single letter in the genetic code.

That was the easy one. Single-gene diseases with well-understood mechanisms: sickle cell, beta-thalassemia, certain forms of blindness, some muscular dystrophies. The therapies work because the problem is simple: one gene, one mutation, one fix. The successes are real, dramatic, and medically revolutionary. They are also the beginning, not the end.

· · ·

The harder applications, the ones that matter for the geopolitical model in this book, involve traits that are controlled by many genes interacting with each other and with the environment. Intelligence. Physical endurance. Disease resistance. Immune function. Cognitive processing speed. These traits are polygenic, meaning they're influenced by hundreds or thousands of genetic variants, each contributing a small effect. Editing one gene doesn't make someone smarter. You'd need to edit hundreds, and you'd need to understand the interactions between them, which we currently don't.

Currently. That word does a lot of work. The gap between what we understand about polygenic traits and what we'd need to understand to engineer them is closing at an accelerating rate, driven by the same AI tools described in Chapter 23. AI can analyze datasets of millions of genomes, identifying the statistical associations between genetic variants and traits with a speed and accuracy that no human research team can match. The UK Biobank, which has sequenced half a million genomes paired with detailed health and trait data, is a goldmine for this kind of analysis. As more genomes are sequenced and more trait data is collected, the statistical power to identify the genetic architecture of complex traits increases.

This doesn't mean someone will be engineering super-geniuses next Tuesday. It means the trajectory is clear: within 10 to 20 years, the understanding of polygenic traits will be sufficient to make meaningful modifications. Not optimization to theoretical maximums. Meaningful shifts: 10 to 20 percent improvements in cognitive function, disease resistance, physical capability. Shifts that are modest individually but decisive when applied across a population and compounded across generations.

• • •

The compounding is what makes genetics different in kind from every other technology in Part Six.

AI doesn't compound. If you build a better AI system, the next generation doesn't inherit the improvement biologically. Each new system has to be built from scratch. Nuclear reactors don't compound. Each one has to be manufactured independently. Longevity treatments don't compound. Each person has to take the drugs or receive the therapy individually.

Genetic modifications to the germline, to the eggs and sperm that produce the next generation, compound. A child born with enhanced disease resistance passes that resistance to their children. Those children pass it to theirs. Each generation starts from a higher baseline than the last. If the modifications include improved cognitive capability, the enhanced generation is better equipped to develop the next round of improvements, which they pass to their children, who are even better equipped. The returns compound the way financial interest compounds: slowly at first, then explosively.

This is why genetics is the most permanent of all the technologies in this book. You can shut down a nuclear reactor. You can turn off an AI system. You can stop taking a longevity drug. You cannot recall a genetic modification once it's been introduced into the human germline and propagated through reproduction. The edit becomes part of the species. Forever.

• • •

He Jiankui's experiment was the proof of concept that the opening of this chapter described. It was poorly designed, inadequately informed, and conducted outside any ethical review framework.

But notice what happened. The technology was used. On humans. By a single researcher with modest resources in a country that officially opposes germline editing. He Jiankui didn't need a Manhattan Project or a billion-dollar facility. He needed a CRISPR kit, a fertility clinic, and willing parents. The barriers to genetic modification of humans are not technological. They are ethical, legal, and regulatory. And those barriers are only as strong as the weakest enforcement regime on the planet.

China officially banned germline editing after the He Jiankui scandal. But China also has the world's largest genomics research infrastructure, the most aggressive national genetics programs, and a government that views biotechnology as a strategic national priority. The ban is real today. Whether it remains real in ten years, when the technology is more precise and the strategic advantages of genetic enhancement become clearer, is a question that the history of arms races answers decisively. When a technology offers a decisive advantage and your adversary might be developing it, you develop it. You may do it quietly. You may do it under a different name. But you develop it.

America has its own regulatory framework through the FDA and NIH that restricts germline editing. These restrictions reflect genuine ethical concerns. They also create a potential strategic vulnerability: if China develops genetic enhancement capabilities while the US maintains its restrictions, the competitive implications play out over generations. An enhanced Chinese population, even modestly enhanced, that passes the enhancement to each subsequent generation, compounds the advantage the way interest compounds in a savings account. Twenty years of enhancement compounding through reproduction produces a population that is measurably different from an unenhanced one.

• • •

The agriculture application is where genetic engineering has the most immediate strategic relevance, and it connects directly to Chapter 14's food crisis.

Nitrogen fixation, the ability to convert atmospheric nitrogen into a biologically usable form, is performed naturally by certain bacteria that live in the root nodules of legumes. Soybeans, clover, alfalfa, and other legumes don't need nitrogen fertilizer because their symbiotic bacteria do the work that the Haber-Bosch process does industrially. If genetic engineering could transfer this capability to cereal crops, rice, wheat, and corn, the dependency on synthetic nitrogen fertilizer would break.

This is the moonshot that Chapter 14 identified as the ultimate solution to the oil-fertilizer-food chain. It's also the hardest genetic engineering problem on the table, because nitrogen fixation involves not a single gene but an entire biological system: the nitrogenase enzyme complex, the oxygen-protection mechanisms that keep the enzyme functional, and the signaling pathways that coordinate the process between the plant and its bacterial symbionts. Transferring the whole system into a cereal crop is a challenge of a different order than editing a single disease gene.

But the tools are converging. AI-powered protein engineering can redesign the nitrogenase complex for function in a different cellular environment. Synthetic biology tools can build the genetic circuits that control the process. CRISPR and its successors can insert the modified system into the crop genome. The timeline is uncertain, maybe 15 years, maybe 30, but the direction is clear: this is a solvable problem, and solving it would be the most consequential agricultural innovation since Haber-Bosch itself.

A world where cereal crops fix their own nitrogen is a world where the Strait of Hormuz doesn't matter for food security. Where natural gas depletion doesn't translate into famine. Where the chain from gas well to dinner plate, which Chapter 14 traced in such alarming detail, simply breaks. That's the potential of genetic engineering applied to the food crisis, and it's worth more than any military system or trade agreement.

. . .

The disease resistance applications are the ones that connect to Chapter 16's pandemic threat.

Some humans are naturally resistant to specific pathogens due to genetic variants they carry. The CCR5-delta32 mutation, the one He Jiankui was targeting, confers resistance to HIV. Other variants provide partial resistance to malaria, certain influenza strains, and other infectious diseases. These are natural experiments: random mutations that happened to provide survival advantages against specific threats.

Genetic engineering could make these resistances universal rather than rare. A population engineered with broad-spectrum pathogen resistance, through modifications to immune function genes, receptor proteins, and inflammatory pathways, would be dramatically less vulnerable to the pandemic scenarios described in Chapter 16. Not invulnerable. Pathogens evolve. But the baseline resilience of the population would be higher, the transmission rates would be lower, and the mortality from any given pandemic would be reduced.

Combined with the mRNA vaccine platforms and AI surveillance systems from Chapters 15 and 22, genetic resistance creates a layered defense against pandemics. The enhanced immune system provides the first line. The rapid vaccine response provides the second. The AI surveillance provides the early warning. No single layer is perfect. Together, they make the species dramatically more resilient to biological threats, whether natural or engineered.

• • •

The ethical questions are real. This book is not going to resolve them, and anyone who tells you the answers are simple is selling you something.

Is it acceptable to modify the human germline, knowing that the changes propagate to future generations who had no voice in the decision? The precautionary principle says no: the risks of unintended consequences are too high and too permanent. The pragmatic argument says the risks of inaction are also high and permanent: if genetic technology offers protection against pandemics, famine, and cognitive decline, refusing to use it condemns future generations to vulnerabilities that could have been prevented.

Who decides what gets modified? If the answer is "parents," then enhancement becomes a consumer product, available to those who can afford it, creating a genetic aristocracy within a generation. If the answer is "governments," then the history of government eugenics programs, from the forced sterilizations of the early 20th century to the racial hygiene theories of the Nazis, provides a cautionary tale that should make anyone's skin crawl.

What counts as "therapy" versus "enhancement"? Curing sickle cell disease is clearly therapy. Increasing cognitive function by 20 percent is clearly enhancement. But what about editing out a gene variant that increases the risk of Alzheimer's by 40 percent? Is that therapy or enhancement? The line between fixing a defect and improving a specification is blurry, and it will get blurrier as the technology advances.

These questions don't have easy answers. But they need to be debated now, before the technology forces the decisions. Because the technology is coming regardless of whether the ethics are resolved. The CRISPR kits are already for sale. The genomes are already being sequenced. The AI systems are already identifying the genetic architecture of complex traits. The only question is whether the deployment happens within an ethical framework or without one.

• • •

Genetics rewrites the code that biology runs on. It's permanent, heritable, and compounding. It addresses the food crisis, the pandemic threat, and the long-term demographic trajectory in ways that no other technology can match. It also raises ethical questions more profound than anything humanity has confronted since the development of nuclear weapons.

The governance vacuum is the most urgent deployment problem. There is no international framework for genetic engineering comparable to the Nuclear Non-Proliferation Treaty or the Chemical Weapons Convention. The closest thing is the Cartagena Protocol on Biosafety, which covers genetically modified organisms in trade but says nothing about human germline editing, nothing about gene drives that could alter entire wild species, and nothing about the military applications of synthetic biology. The He Jiankui case demonstrated that a single rogue researcher in one country can alter the human germline with equipment that costs less than a used car. No treaty prevents this. No enforcement mechanism exists to stop it. The barriers are national laws that vary wildly in their scope and enforcement, and scientific norms that work only as long as everyone agrees to follow them.

What is needed is an international biosafety regime with actual teeth: mandatory registration of all germline editing experiments with an international body, an inspection mechanism modeled on the IAEA, export controls on the most dangerous synthetic biology tools, and agreed red lines on human enhancement research that would give one nation a military advantage. This regime needs to be negotiated now, while the technology is still young enough that norms can be established before capabilities outrun the rules. The Nuclear Non-Proliferation Treaty was signed in 1968, when only five countries had the bomb. If the international community had waited until twenty countries had it, the treaty would never have been possible. The same logic applies to genetic engineering. The window for establishing governance is open now. It will not stay open forever.

But genetics operates at the speed of reproduction: a generation per iteration. There is a technology that operates at the speed of surgery, adding hardware to the building while people are still living in it.

Chapter 26: Ghost in the Shell

In January 2024, a 29-year-old quadriplegic named Noland Arbaugh played chess using nothing but his thoughts. A chip implanted in his motor cortex read his neural signals, translated them into digital commands, and moved the cursor on a screen. He had been paralyzed from the shoulders down since a diving accident. Now he was browsing the internet, streaming music, and controlling a computer without touching anything. He described the experience as using the Force. The bandwidth was low. The capabilities were limited. The surgery was invasive. But the principle was proven: a machine can read human thought and translate it into action. That was 2024. This chapter is about where it goes from here.

In the 1995 anime film *Ghost in the Shell*, the protagonist, Major Motoko Kusanagi, is a human consciousness inhabiting a fully synthetic body. Her brain is augmented with cybernetic implants that allow direct neural interface with computer networks. She can download skills, access databases by thinking about them, communicate telepathically with other augmented individuals, and perceive the world through electronic senses that exceed anything biological eyes and ears can provide. Her body is stronger, faster, and more durable than any natural human's. She is, in every functional sense, a different kind of being than the unaugmented humans she works alongside.

The film asks a philosophical question: if you replace enough of the biological with the technological, is the "ghost," the consciousness, the self, still human? It's a good question. It was science fiction in 1995. It is becoming an engineering question in 2026.

This chapter is about the technologies that are turning Kusanagi from a cartoon into a blueprint. And about what happens to the geopolitical model when some humans are fundamentally more capable than others, not through training or education or genetics, but through hardware.

• • •

The progression has already started, and most people haven't noticed because the early steps look medical rather than military.

Cochlear implants have given hearing to over a million deaf people worldwide. The device bypasses the damaged biological hearing system and feeds electrical signals directly to the auditory nerve. The brain learns to interpret those signals as sound. A human being is hearing through a machine interface with a nerve. That's a neural implant. It's been routine for decades.

Deep brain stimulation uses electrodes implanted in the brain to treat Parkinson's disease, essential tremor, and severe depression. The electrodes deliver electrical pulses that modulate the neural circuits responsible for the symptoms. Hundreds of thousands of people are walking around with wires in their brains that modify how their neurons fire. That's a brain-computer interface. It's been FDA-approved since 1997.

Neuralink's Arbaugh implant, described at the opening of this chapter, proved the principle. The device, called Telepathy, reads neural signals from the motor cortex and converts them into digital commands. The current version is crude by the standards of what is coming. But every transformational technology starts crude.

Each of these is a medical device treating a disability. None of them are enhancement in the strategic sense. But the trajectory from treating disability to augmenting capability is a straight line, and every medical advance moves the starting point closer to the military and economic applications that matter for this book.

• • •

The near-term enhancement technologies, available within the next 10 to 15 years, fall into three categories.

Sensory augmentation. Current human senses operate within narrow biological bands. We see a tiny slice of the electromagnetic spectrum. We hear frequencies between roughly 20 Hz and 20,000 Hz. We can't detect magnetic fields, radioactivity, chemical traces below certain concentrations, or infrared heat signatures. Each of these limitations can be bypassed with technology that feeds non-biological sensory data directly to the nervous system.

A soldier with infrared vision implants sees in the dark without goggles. A pilot with accelerometer implants feels the aircraft's orientation through their nervous system rather than reading instruments. A submarine technician with ultrasonic hearing detects mechanical anomalies that no biological ear can perceive. Each of these is an extension of existing technology: the sensors already exist, the neural interfaces are being developed, and the military applications are obvious enough that DARPA has been funding research in this area for years.

Cognitive augmentation. The Neuralink-style brain-computer interface is the first step toward something far more significant than cursor control. If you can read neural signals well enough to move a cursor, you can eventually read them well enough to detect intentions, recall states, and cognitive patterns. And if you can write signals back into the brain, not just read them out, you can provide information directly to the cortex without going through the eyes and ears.

Imagine a fighter pilot who receives threat data directly in their visual cortex, overlaid on their natural vision without a helmet display. A intelligence analyst who queries a database by thinking the question and receives the answer as if they'd remembered it. A surgeon who accesses a patient's complete medical history and surgical reference material without looking away from the operating field. Each of these requires a higher-bandwidth brain-computer interface than currently exists, but the engineering roadmap is clear: more electrodes, better signal processing, bidirectional communication, and AI-mediated translation between neural patterns and digital information.

Physical augmentation. Powered exoskeletons are already in use for industrial applications and are being developed for military use. They amplify human strength, allowing a person to lift heavy loads, carry equipment over long distances, and operate in conditions that would exhaust an unaugmented body. The current generation is external, worn like a suit. Future generations will integrate more closely with the body, potentially including implanted actuators, reinforced skeletal structures, and synthetic muscle fibers that work alongside biological ones.

The military applications are obvious: a soldier in a powered exoskeleton carries more ammunition, moves faster under load, and fights longer than an unaugmented one. The shipyard applications from Chapter 3 are equally direct: a welder in an exoskeleton works longer shifts, lifts heavier components, and maintains precision in positions that would fatigue a biological body in minutes. The demographic crisis meets its technological answer: fewer workers, but each worker is physically capable of doing the work of two or three.

• • •

The longer-term technologies, 15 to 30 years out, are the ones that start to resemble Kusanagi more than a medical patient.

Direct neural interfaces with AI systems. Not the crude cursor-control of current BCIs, but high-bandwidth connections that allow a human brain to interact with AI at something approaching the speed of thought. A person with this interface doesn't use AI as a tool. They think with it. The AI becomes an extension of their cognitive process, handling the computational tasks that biological neurons are slow at, mathematical modeling, pattern recognition in large datasets, language translation, while the biological brain handles the tasks it's good at: creative insight, emotional judgment, ethical reasoning, strategic intuition.

The resulting entity is neither human nor AI. It's something new: a hybrid intelligence that combines the strengths of both architectures. The human provides the consciousness, the values, the goals. The AI provides the processing power, the data access, the computational speed. Together they can do things that neither could do alone. That's the ghost in the shell: a human consciousness operating through a technologically augmented substrate that extends its capabilities far beyond biological limits.

Synthetic organ replacement. If a heart can be replaced with a mechanical pump, and it can, then the principle extends to every organ in the body. Synthetic lungs that extract oxygen more efficiently. Synthetic kidneys that filter waste without deteriorating. Synthetic liver systems that process toxins without accumulating damage. Each replacement removes a point of biological failure. A person with synthetic organs doesn't get heart disease, doesn't get kidney failure, doesn't get liver cirrhosis. The failure modes that currently kill people in their 60s and 70s are simply removed from the equation.

Combined with the longevity interventions from Chapter 24, which address the cellular and molecular aspects of aging, synthetic organ replacement addresses the mechanical aspects. The cells stay young. The organs don't wear out. The productive lifespan extends not by 15 or 20 years but potentially by 50 or more. The demographic model doesn't just improve. It transforms into something unrecognizable from the current framework.

. . .

The geopolitical implications are the ones that keep strategic planners awake.

An enhanced military force is not just incrementally better than an unenhanced one. It's categorically different. A fighter pilot with direct neural interface to their aircraft's systems reacts at machine speed instead of human speed. A special operations team with integrated communications doesn't need radio equipment that can be jammed or intercepted. An intelligence analyst with AI cognitive augmentation processes in hours what currently takes a team weeks. A submarine crew with enhanced endurance operates without the fatigue that degrades human performance on long patrols.

The country that fields enhanced military personnel first gains an advantage that training and equipment cannot match. You can't train a baseline human to react at millisecond speed. You can't equip a biological soldier with infrared vision that's integrated into their nervous system rather than strapped to their face. The enhancement IS the advantage, and it's not something the other side can counter without developing the same capability.

This creates an enhancement arms race that has the same structural dynamics as the nuclear arms race, with one critical difference: nuclear weapons are deterrents, useful primarily because they're never used. Enhancement technology is useful precisely because it is used, every day, in every military operation. The incentive to develop it is stronger than the incentive to develop nuclear weapons, because the payoff is immediate and continuous rather than theoretical and catastrophic.

. . .

The divergence problem is the one that *The Birth of the Augmented Human*, the third book in this series, explores in depth. This chapter will sketch it briefly because it connects to the geopolitical model.

Enhancement technology will be expensive initially. The first neural interfaces, the first synthetic organs, the first integrated AI cognitive systems will be available to wealthy nations and wealthy individuals within those nations. The gap between enhanced and unenhanced humans will be, at first, a gap between rich and poor, between developed and developing countries, between the first adopters and everyone else.

Over time, as the technology matures and costs decline, the gap may narrow. Cochlear implants went from exotic to routine over two decades. Smartphones went from luxury to ubiquity in ten years. The historical pattern suggests that breakthrough technologies eventually democratize. But "eventually" can mean decades, and during those decades the early adopters compound their advantage.

The scenario that keeps ethicists awake is the one where enhancement doesn't democratize fast enough. A world where the wealthy nations and the wealthy individuals within them have enhanced cognitive capability, extended lifespans, superior physical function, and integrated AI assistance, while the rest of humanity operates on the biological hardware evolution provided. Two tiers of human being, separated not by race or nationality or ideology but by access to technology.

The enhanced tier pulling further ahead every year because their cognitive advantages allow them to develop the next generation of enhancements faster, creating a feedback loop that the unenhanced tier cannot enter.

That's not dystopian fiction. That's the logical extrapolation of current technology trajectories and current economic structures. Whether it happens depends on whether societies choose to make enhancement a public good rather than a private luxury. And that choice depends on the political systems that Chapter 20 described being captured by extraction interests and Chapter 21 described being staffed by a population that can't think critically about the choices being made on its behalf.

• • •

This is where all the threads of Part Six come together.

But deployment faces a regulatory wall. The FDA's current framework was designed to approve drugs and medical devices intended to treat disease. Enhancement, making a healthy person function beyond normal human parameters, does not fit neatly into any existing regulatory category. A brain-computer interface that restores motor control to a quadriplegic is a medical device treating a disability. The same interface in a healthy soldier that allows direct neural control of a drone swarm is something the regulatory system has no box for. The Neuralink device Noland Arbaugh received was approved under the FDA's investigational device exemption for clinical trials, a pathway designed for experimental treatments of medical conditions. There is no equivalent pathway for enhancement technologies. Until the FDA either creates one or Congress mandates one, every enhancement technology will have to be laundered through a medical application first: the device treats a disease, and the enhancement is an off-label side effect. That is inefficient, dishonest, and guaranteed to ensure that enhancement technology develops faster in countries with less scrupulous regulatory systems.

The military offers a partial workaround. DARPA has been funding neural interface and enhancement research for years, and military applications do not require FDA approval in the same way commercial products do. A soldier who volunteers for an experimental neural interface is operating under military medical authority, not civilian regulatory authority. This means the first enhanced humans in America will almost certainly be military personnel, just as the first users of GPS, the internet, and night-vision technology were military personnel. The technology will migrate to civilian applications later, once the safety data from military use provides the evidence base that the FDA needs to create an approval pathway. This is not ideal. It creates a two-tier system where soldiers are the guinea pigs and civilians wait for the results. But it is the most realistic path given the current regulatory landscape, and it has the advantage of aligning the technology development with the strategic imperative that makes it most urgent.

The PE extraction from Chapter 20 hollows out the institutional capacity to develop and deploy enhancement technology equitably. The education crisis from Chapter 21 means fewer people capable of building the technology in the first place. Nuclear energy from Chapter 22 provides the power to run the computational and manufacturing systems that enhancement requires. AI from Chapter 23 provides the intelligence that makes brain-computer interfaces functional. Longevity from Chapter 24 provides the biological foundation that enhancement builds on. Genetics from Chapter 25 provides the germline modifications that compound across generations.

Each technology amplifies the others. AI makes nuclear deployment faster. Nuclear powers the AI data centers. Longevity extends the productive lives of the scientists developing AI. Genetics creates a baseline population that's healthier and more cognitively capable. Enhancement integrates all of it into a human organism that is qualitatively different from the one that evolution produced.

The country that deploys this full stack first, nuclear energy feeding AI systems feeding longevity research feeding genetic engineering feeding enhancement technology, doesn't just have a competitive advantage. It has a civilizational advantage that compounds across generations and across every domain at once.

That's the prize. But none of it happens without the ecosystem that produces the breakthroughs and the domain that makes the resources infinite. The innovation engine is the ecosystem. Space is the domain. And both are under threat.

Chapter 27: The Innovation Thesis

Every technology described in the preceding five chapters, nuclear energy, AI, longevity science, genetic engineering, and human enhancement, was either invented or substantially advanced in the United States. The pressurized water reactor was American. The transistor was American. The internet was American. CRISPR was characterized by American and European researchers. The mRNA vaccine platform was developed by American companies. The first brain-computer interface company to implant a human was American. The leading AI research labs are American.

This is not patriotic chest-thumping. It's a data point that matters for the argument. America has been the world's most prolific generator of breakthrough innovation for over a century, and the structural reasons for that are the single most important strategic advantage the country possesses. More important than the Navy. More important than the nuclear arsenal. More important than the dollar's reserve currency status. Because all of those can be outbuilt, outproduced, or outmaneuvered. Innovation cannot be outmaneuvered. You either produce breakthroughs or you don't. And the United States produces more of them than any other country on Earth.

The question is why. And the answer matters because the "why" is under threat from every direction, and if it breaks, the entire optimistic case breaks with it.

• • •

The American innovation engine runs on four inputs that no other country combines at the same scale.

First, research universities. The top 50 research universities in the world include roughly 30 American institutions. These aren't just teaching schools. They're research laboratories where fundamental science happens, where breakthrough discoveries are made, and where the graduate students and postdoctoral researchers who will start the next generation of companies are trained. The federal research funding system, primarily through the NIH, NSF, DOE, and DARPA, channels tens of billions of dollars into university-based research every year.

DARPA deserves special mention because it is the institutional model that every other country tries to copy and none has replicated. The Defense Advanced Research Projects Agency funds high-risk, high-reward research with the explicit goal of producing technological surprises. The internet started as a DARPA project. GPS started as a DARPA project. Stealth technology started as a DARPA project. The mRNA research that produced COVID vaccines received DARPA funding. The agency's model of giving smart people money and getting out of their way has produced more strategically consequential innovations per dollar than any other institution in history.

The model is so effective that Congress has created civilian versions: ARPA-E for energy research, established in 2009 to fund breakthrough energy technologies that the private sector considers too risky, and ARPA-H for health research, created in 2022 to accelerate biomedical breakthroughs with a $2.5 billion initial budget. Both are modeled on DARPA's program manager structure, where individual researchers with deep domain expertise are given significant budgets and minimal bureaucracy for three-to-five-year terms, then rotated out to prevent institutional calcification. The model works because it attracts people who want to change their field, gives them the resources to do it, and then forces them to leave before they become administrators.

Second, venture capital. The US has the deepest, most experienced, and most aggressive venture capital ecosystem on the planet. VC doesn't just provide money. It provides the institutional knowledge to turn a laboratory discovery into a commercial product. How to build a team. How to find a market. How to scale production. How to handle regulation. The combination of federally funded basic research feeding into VC-funded commercialization is a pipeline that converts scientific discoveries into economic and strategic power faster than any competing model.

Third, immigration. Chapter 12 made the demographic case. The innovation case is equally strong. According to the National Foundation for American Policy, immigrants have founded 55 percent of America's startup companies valued at $1 billion or more---319 out of 582 companies as of their 2022 study. The co-founders of Google, Tesla, Yahoo, eBay, and dozens of other world-changing companies came from outside the United States. Immigrants hold key leadership positions in 71 percent of billion-dollar startups. They came because American universities were the best, American venture capital was the most available, and the American market was the largest. That magnetic pull, the ability to attract the most ambitious and talented people from every country on Earth and give them the resources to build things, is an innovation advantage that no closed society can replicate.

Fourth, a culture that tolerates failure. This sounds soft, but it's structurally decisive. In most countries, business failure carries social stigma that discourages risk-taking. In the United States, a failed startup is a resume line, not a scarlet letter. The willingness to fund ten companies knowing that eight will fail and one might change the world is a cultural trait that the venture capital system institutionalized and that the broader society accepts. Risk tolerance is the fuel that the other three inputs burn. Without it, the universities produce papers instead of products, the capital sits in bonds instead of startups, and the immigrants go somewhere else.

• • •

China has tried to replicate this model and has succeeded in some dimensions while failing in others. Understanding where China succeeds and where it doesn't is essential to assessing the innovation competition.

China's strengths in innovation are real and growing. Massive state investment in research and development. A university system that graduates millions of STEM students. A manufacturing base that can scale production faster than any other country. An internal market of 1.4 billion people that provides a testing ground for new technologies. And a government that can direct resources toward strategic technologies without the market inefficiencies and political interference that slow American deployment.

China excels at what might be called "innovation by application." Taking technologies invented elsewhere and deploying them faster, cheaper, and at larger scale. Chinese solar panel manufacturing didn't invent the solar cell. It scaled it. Chinese battery technology didn't invent the lithium-ion battery. It mass-produced it. Chinese high-speed rail didn't invent the bullet train. It built more of it than the rest of the world combined. The ability to take a working technology and deploy it at national scale in five years is an enormous competitive advantage, and it's one the US consistently fails to match.

Where China struggles is at the frontier. The breakthrough discoveries, the fundamental science, the genuinely novel ideas that create entirely new technological domains. China produces more scientific papers than any other country, but citation analysis consistently shows that the highest-impact research still comes disproportionately from American and European labs. The Chinese system rewards volume over originality, safe incremental work over risky breakthrough attempts. The researchers are smart and well-funded. The institutional culture discourages the kind of radical creativity that produces paradigm shifts.

The authoritarian political system is the deeper constraint. Innovation at the frontier requires intellectual freedom: the ability to challenge existing theories, pursue unfashionable ideas, publish results that embarrass the establishment, and fail repeatedly without career consequences. A system that monitors speech, punishes dissent, and rewards conformity is structurally hostile to the kind of thinking that produces breakthroughs. China can optimize. It struggles to originate. And in the technology domains that matter most for this book, origination is what counts.

• • •

The American innovation engine is under threat, and the threats are mostly self-inflicted.

Federal research funding has been flat in real terms for years and is under active political threat. The current administration has proposed cuts to the NIH, NSF, and DOE that would reduce the basic research pipeline at exactly the moment when the technologies described in this book need more investment, not less. DARPA's budget is a rounding error compared to the defense procurement budget, and it's politically vulnerable because its projects are inherently risky and some of them fail, which makes for easy targets in budget hearings.

The university system, as Chapter 21 described, is financialized and bureaucratized to a degree that threatens research quality. Faculty spend more time writing grant proposals than doing research. Administrative overhead consumes a growing share of university budgets. The publish-or-perish culture rewards incremental papers over breakthrough research. The tenure system, originally designed to protect academic freedom, increasingly selects for conformity rather than originality. The best researchers are still world-class. The system that supports them is getting worse.

Immigration restrictions threaten the talent pipeline. The H-1B visa system is capped at levels set in the 1990s, before the tech industry exploded. Wait times for employment-based green cards for Indian and Chinese nationals, who constitute a huge share of the STEM talent pipeline, stretch to decades. A Chinese AI researcher who gets a PhD at Stanford and wants to stay in the US to found a company may have to wait fifteen years for permanent residency, during which time they can't change employers freely or take the entrepreneurial risks that the innovation system depends on. Canada, the UK, and Singapore have built fast-track immigration programs explicitly designed to poach the talent that American immigration policy drives away.

And the venture capital ecosystem, while still the strongest in the world, is showing signs of the same short-termism that afflicts every other American financial institution. The average VC fund has a ten-year lifespan. Nuclear energy, longevity science, genetic engineering, and space technology all have development timelines that exceed ten years. What results is a VC system that over-invests in software companies that can return capital in five years and under-invests in the deep-tech companies whose products take fifteen years to develop but could change civilization.

The technologies described here are exactly the ones that the VC model is worst at funding. Nuclear startups cannot raise venture capital because the development timeline exceeds the fund's life. Longevity biotech gets funded for drug discovery, which can return capital in five to seven years, but not for the fundamental aging research that would produce the breakthroughs described in Chapter 24. Space companies get funded for satellite internet, revenue in three years, but not for asteroid mining, revenue in twenty.

The innovation engine is optimized for incremental improvement in software and services. It is systematically starving the civilizational-scale technologies that Part Six describes. This is the innovation version of the missile production gap from Chapter 2: a system optimized for peacetime that cannot surge for the strategic moment.

• • •

The innovation thesis is that the United States retains a structural advantage in producing the breakthrough technologies that define the strategic competition. The advantage is real, measurable, and not easily replicated. But it is narrower than it was a generation ago, more fragile than the technology sector acknowledges, and under threat from policy decisions that could be reversed if anyone in a position of authority understood what was at stake.

The policy prescription is clear. Double federal research funding, particularly through DARPA and its civilian equivalent ARPA models. Reform immigration to create a fast-track system for STEM talent that makes it easier to stay in the US than to leave. Protect university research from bureaucratic overhead and political interference. Create long-duration funding mechanisms for deep-tech development that the VC system can't fund. And stop cutting the seed corn: the basic research that produces breakthroughs twenty years from now is the research that gets funded, or doesn't get funded, today.

None of this is expensive by federal budget standards. The entire NSF budget is roughly $10 billion, less than the cost of a single aircraft carrier. The entire DARPA budget is roughly $4 billion, less than the cost overrun on the Ford-class carrier program. The research funding that produces the innovations that maintain American strategic supremacy costs less than the military hardware that buys time until the innovations arrive. And unlike the hardware, which wears out and gets sunk, the innovations compound.

The country that leads in innovation leads in everything. The country that falls behind in innovation falls behind in everything. There is no military, economic, or diplomatic substitute for being the place where the future gets invented. America has been that place for a century. It can remain that place. But it requires treating the innovation pipeline as what it is: the foundation of national power, not a budget line to be trimmed when Congress needs to fund a tax cut.

• • •

Part Six has described the hollowed-out country and the tools to rebuild it. The extraction model from Chapter 20 and the education crisis from Chapter 21 are the damage report. Nuclear energy, AI, longevity, genetics, enhancement, and the innovation ecosystem are the repair kit. Each tool addresses specific crises from Parts One through Five. Together, they describe a path from the fragile, overextended, under-capacity America of 2026 to a fundamentally more capable, more resilient, and more competitive civilization.

But there's one more domain that hasn't been discussed. One that sits above the map, literally, and that could amplify or negate everything in the preceding chapters depending on who controls it.

The high ground. The place where satellites orbit and missiles fly and the next century's most consequential competition may already be underway.

Chapter 28: The High Ground

In every military engagement since humans started fighting on terrain more complex than a flat field, the force that holds the high ground has the advantage. You can see farther. You can shoot downhill. The enemy has to climb to reach you. From the walls of Troy to the ridgeline at Gettysburg to the mountains of Afghanistan, elevation is power.

Space is the ultimate high ground. And modern civilization depends on it far more completely than most people understand.

. . .

GPS, the Global Positioning System, is a constellation of 31 satellites orbiting at roughly 12,550 miles above the Earth. Every smartphone, every aircraft navigation system, every ship's chart plotter, every precision-guided munition, every agricultural GPS system that drives a tractor in straight lines across a field, depends on signals from those satellites. GPS is so embedded in the infrastructure of daily life that most people don't think about it. It's just there, like electricity. You turn on your phone and it knows where you are.

If GPS went dark, the consequences would cascade through every system described here. Military operations would revert to map-and-compass navigation. Precision-guided munitions would become unguided. Commercial aviation would lose its primary navigation system. Container ships would lose their positioning. The financial system, which uses GPS timing signals to synchronize transactions, would begin malfunctioning within hours. The precision agriculture systems from Chapter 23 would stop functioning. Emergency services would lose their dispatch systems.

That's one constellation. Now add the communications satellites that carry global internet traffic, telephone calls, and television signals. The weather satellites that provide the forecast data every military operation and agricultural decision depends on. The reconnaissance satellites that provide the intelligence pictures that drive foreign policy. The early warning satellites that detect missile launches and prevent nuclear miscalculation. Every one of these is a machine in orbit, fragile, predictable in its path, and increasingly targetable.

America has more assets in space than any other country. It also has more to lose.

• • •

China demonstrated the ability to destroy satellites in 2007 when it shot down one of its own weather satellites with a ground-launched missile. Russia demonstrated the same capability in 2021, destroying a defunct Soviet-era satellite. Both tests produced massive clouds of debris that will orbit for decades, threatening every other object in the same altitude range. The tests were strategic messages: we can blind you.

Anti-satellite weapons come in several varieties. Direct-ascent missiles, like the ones China and Russia tested, physically destroy the satellite. Directed energy weapons, ground-based lasers, can blind or damage satellite sensors without creating debris. Co-orbital weapons, small satellites that maneuver close to a target and disable or destroy it, are harder to attribute and harder to detect. Electronic warfare can jam satellite signals or spoof them, making a GPS receiver think it's somewhere it isn't. Cyber attacks can compromise the ground stations that control satellite constellations, turning the satellites into expensive space junk.

The vulnerability is asymmetric. The US depends on space more than any potential adversary. China has BeiDou as a GPS alternative and is less dependent on space-based communications because its military operates primarily within its own region. Russia has GLONASS. Both countries would be hurt by the loss of space assets, but they would be hurt less than the United States, whose entire military doctrine, from precision strike to intelligence collection to command and control, is built on space-based systems.

A conflict that begins with the destruction of each side's satellite constellations favors the side that is less dependent on those constellations. That's China. The US military, blinded and deafened by the loss of its space architecture, would have to fight the way it fought in World War II: with maps, radio, visual reconnaissance, and unguided weapons. It hasn't trained for that in decades. The institutional muscle memory is gone. The weapons aren't designed for it. The entire force would be degraded to a fraction of its capability overnight.

• • •

Kessler syndrome is the nightmare scenario that connects the military vulnerability to something worse.

In 1978, NASA scientist Donald Kessler proposed that a collision between objects in orbit could create a chain reaction: the debris from the collision hits other objects, creating more debris, which hits more objects, in an exponentially growing cascade that eventually fills an orbital band with so much junk that no satellite can survive there. The specific altitude doesn't matter. What matters is that the cascade, once started, is self-sustaining and irreversible on any human timescale. The debris eventually decays and falls back to Earth, but "eventually" means decades to centuries depending on the altitude.

The 2007 Chinese anti-satellite test produced over 3,500 pieces of trackable debris and an estimated hundreds of thousands of pieces too small to track but large enough to destroy a satellite. Each piece is orbiting at roughly 17,000 miles per hour. At that speed, a paint chip carries the energy of a bullet. A bolt carries the energy of a hand grenade. The debris will be a hazard for generations.

Now imagine a conflict in which both sides destroy each other's satellites. Dozens or hundreds of satellite kills, each producing thousands of debris fragments, in the most populated orbital bands. The cascade begins. Low Earth orbit, where most reconnaissance and communication satellites operate, becomes impassable. GPS satellites in medium Earth orbit are less vulnerable but not immune. Geostationary satellites, the ones that provide communications and weather coverage, orbit high enough that the debris risk is lower but not zero.

A full Kessler cascade would not just blind the militaries. It would end the space age for generations. No new satellites could be launched into the affected orbits because they would be destroyed by debris within days. No satellite repairs, no resupply, no expansion. Humanity would be locked out of the orbital bands it depends on for communications, navigation, weather forecasting, and scientific observation. The economic and civilizational impact would dwarf any military outcome of the conflict that caused it.

This is not a theoretical concern. The density of objects in low Earth orbit is already approaching levels where some scientists believe the cascade could be triggered by a single unlucky collision, even without a military conflict. The mega-constellations being launched by SpaceX, Amazon, and others are adding thousands of objects to an orbital environment that is increasingly crowded. Every launch adds risk. Every defunct satellite that isn't deorbited is a potential trigger.

The strategic response to space vulnerability has two dimensions, and the US is pursuing both, though not fast enough.

Resilience through proliferation. Instead of relying on a small number of large, expensive, exquisite satellites, build constellations of hundreds or thousands of small, cheap, replaceable ones. If the adversary destroys ten, you have five hundred more. The Starlink constellation, originally built for commercial internet service, has demonstrated military utility in Ukraine, providing communications that Russian electronic warfare couldn't suppress because there were too many satellites to jam at the same time. The Pentagon's Space Development Agency is pursuing a similar approach for military applications: a proliferated constellation of small satellites in low Earth orbit that provides communications, missile tracking, and data relay.

The logic is the same cost-exchange ratio that Chapters 1 and 2 described for missiles: make your assets cheap and numerous so that the cost of destroying them exceeds the cost of replacing them. A $100 million satellite that takes five years to build is a target. A $500,000 satellite that takes six months to build is a rounding error. Destroy it and another one launches next month.

Offensive counter-space capability. If the adversary can blind you, you need the ability to blind them. The US Space Force, established in 2019, is developing the doctrine and the weapons for space warfare. The details are mostly classified, but the general categories are known: ground-based lasers to dazzle satellite sensors, electronic warfare systems to jam satellite communications, cyber tools to compromise satellite control systems, and likely kinetic weapons for destroying satellites in extremis.

The tension between offensive counter-space capability and the Kessler risk is obvious. Every satellite you destroy creates debris that threatens your own satellites. The logic of space warfare tends toward mutual blinding followed by mutual inability to use space, which as noted above is an outcome that hurts the more space-dependent side more. The US needs to be prepared for space warfare while at the same time working to prevent it, which is the same paradox that governs nuclear weapons: you maintain the capability so you never have to use it.

The Space Force itself is still finding its institutional footing. Established in December 2019 as the sixth branch of the US military, it inherited roughly 16,000 active-duty personnel and a $15 billion budget from the Air Force's space operations. Its mission is to protect American and allied interests in space, deter aggression in and from space, and conduct space operations. But it is the smallest military branch by far, its budget is dwarfed by the other services, and it is fighting for bureaucratic survival in a Pentagon culture that still thinks of space as a support function rather than a warfighting domain. The Space Development Agency, which is building the proliferated constellation of small satellites, was only transferred into the Space Force in 2023. The organizational plumbing is still being connected. The doctrine is still being written. The weapons programs are still mostly in development. Meanwhile, China launched more orbital missions in 2024 than any other country, and its military space program is not struggling with institutional identity crises. It is building capability with the same industrial urgency that the PLAN brings to shipbuilding.

• • •

Beyond the military domain, space is the arena where the resource constraints could eventually be broken entirely.

Asteroid mining is not science fiction. It's a business plan. Near-Earth asteroids contain concentrations of metals, iron, nickel, cobalt, platinum group elements, that dwarf anything available in terrestrial deposits. A single metallic asteroid a kilometer in diameter could contain more platinum than has ever been mined in human history. The rare earth elements that Chapter 18 described China dominating exist in quantities in asteroids that would make Chinese processing monopolies irrelevant.

The technical challenges are real. Reaching an asteroid, mining it in microgravity, and returning the materials to Earth or processing them in space requires propulsion, robotics, and materials science that are at the edge of current capability. The cost of the first missions will be enormous. But the costs of space access are falling dramatically, driven by SpaceX's reusable rocket technology and the competition it's inspired. What cost a billion dollars a decade ago costs tens of millions today. What costs tens of millions today will cost millions in another decade.

The first companies are already flying. AstroForge, a California startup founded in 2022, launched the first commercial deep-space asteroid prospecting mission in February 2025, targeting a metallic near-Earth asteroid for a flyby at a total mission cost of $6.5 million. The spacecraft was lost to communication failures, but the company is planning a landing mission for 2026 and has raised $55 million to date.

TransAstra, also in California, is developing a different approach: using concentrated sunlight to extract water from carbonaceous asteroids for use as in-space rocket propellant. Two previous asteroid mining companies, Planetary Resources and Deep Space Industries, collapsed in the 2010s because launch costs were still too high. The economics have changed. AstroForge built its spacecraft in nine months for $3.5 million. A decade ago, a comparable mission would have cost hundreds of millions.

The timeline for commercially viable asteroid mining is probably 20 to 30 years. Within the planning horizon. And when it arrives, it changes the resource equation that underlies every geopolitical competition described in Parts Four and Five. If rare earths can be mined from asteroids, China's processing monopoly is worthless. If phosphate can be extracted from carbonaceous asteroids, Morocco's reserves are irrelevant. If iron and nickel are available in space, the terrestrial mining constraints that limit industrial capacity become historical curiosities.

The country that controls space-based resource extraction has solved the resource problem the way nuclear energy solves the power problem: by accessing a supply so vast that the concept of scarcity ceases to apply. That's not a near-term solution. It's a generational one. But it's the endgame for the resource competition, and the country that invests in it now will dominate it later.

• • •

Space-based solar power is the other long-term play that deserves mention. A solar collector in geostationary orbit receives sunlight 24 hours a day with no atmospheric interference. The energy is converted to microwaves and beamed to a receiving station on Earth. The concept has been studied since the 1970s. The physics work. The engineering is achievable with current or near-term technology. The cost is falling as launch costs fall.

Space-based solar is baseload power from the sun. It doesn't need batteries. It doesn't have intermittency. It works at night and through clouds. A constellation of solar power satellites could provide clean, continuous, unlimited energy to any point on the Earth's surface. Combined with the nuclear energy described in Chapter 22, it creates an energy system that is genuinely inexhaustible and genuinely independent of every terrestrial resource constraint and geopolitical chokepoint.

That's the real strategic significance of space, beyond the satellites and the weapons and the orbital mechanics. Space is where the resource and energy constraints that drive every conflict in this book stop being constraints. It's where the zero-sum competition for finite terrestrial resources becomes a positive-sum expansion into effectively infinite resources. And the country that gets there first doesn't just win the competition. It transcends it.

• • •

Part Six has described nine tools: nuclear energy, AI, longevity science, genetic engineering, human enhancement, the innovation thesis, and now space as a domain. Two of those, PE extraction and education, are problems to solve. Seven are capabilities to build. Together they describe a civilization that is qualitatively different from the one that exists today, one that has solved its energy crisis, extended its productive lifespan, enhanced its cognitive and physical capabilities, broken its resource dependencies, and expanded into a domain that makes terrestrial scarcity obsolete.

That civilization is possible. Every technology described in Part Six exists or is under active development. None require miracles. All require decisions.

The question that remains is the one the whole book has been building toward: in what kind of world do these tools get deployed? A world of orderly competition between two hemispheric powers? A world of chaotic fragmentation? A world of managed decline or renewed ascent?

Part Seven answers that question. And it starts with the hardest version of it: if the world splits in two, which half would you rather live in?

Part Seven: Stuck in the Middle

Chapter 29: Two Hemispheres, Stress-Tested

This book has described a world that is fracturing. The American-led order built after 1945 is breaking apart under the combined pressure of strategic withdrawal, demographic decline, resource depletion, climate change, and a multi-theater conflict that is consuming the military capacity of the country that built the order. Meanwhile, China is assembling an alternative system through economic infrastructure, resource acquisition, and strategic patience, positioning itself to dominate the Eastern Hemisphere the way the United States has dominated the Western one for two centuries.

The model that emerges from all of this, the one this book proposes as the most stable available configuration for the next fifty years, is a bipolar world. Two hemispheres. Two dominant powers. Two spheres of influence. Not the Cold War's ideological confrontation, not a global struggle for the soul of humanity, but a practical division of the planet along geographic lines that reflect the structural advantages each power holds.

This chapter stress-tests that model against everything the book has described. Because a model that sounds neat in a think tank briefing has to survive contact with the four horsemen, the quiet empire, the hollowed-out country, and the wars that are already underway. If it can't survive that contact, it's useless.

. . .

The Western Hemisphere bloc. Start with what it has.

The United States, with the world's largest economy, most capable military, most productive innovation system, most favorable geography, and the demographic advantage of immigration. Canada, with vast natural resources, freshwater, arable land expanding under climate change, and deep integration with the American economy. Mexico, with a young and growing population, a manufacturing base that nearshoring is expanding rapidly, and a geographic position that makes it the natural partner for any American industrial strategy that requires cheap labor close to home.

Brazil, with the largest economy in South America, enormous agricultural capacity, massive freshwater reserves in the Amazon basin, and a young population. Colombia, Chile, Peru, Argentina, each with specific resources and capabilities that complement the bloc. The Caribbean, Central America, smaller in economic terms but geographically essential for control of the maritime approaches.

The hemisphere's combined resource base is staggering. Energy: oil from Venezuela, the US, Canada, Brazil, and Mexico. Natural gas from the US, Canada, Argentina, and Trinidad. Uranium from Canada and the US. Solar and wind resources across the tropics and plains. Agricultural land: the most productive farmland on the planet, from the American Midwest through the Brazilian cerrado to the Argentine pampas. The Mississippi system. The Amazon. The Great Lakes. Water, energy, food, and minerals sufficient to sustain a civilization of a billion people indefinitely.

The hemisphere's demographic profile is the healthiest in the developed world. The US maintains population growth through immigration. Mexico and Central America provide a young labor force. Brazil's population is still growing, though slowing. The hemisphere doesn't face the demographic cliff that China, Japan, South Korea, and Europe are staring at. It faces a gentler slope, manageable with the technology described in Part Six.

• • •

Now stress-test it.

Energy. The hemisphere is energy-self-sufficient and then some. Under the nuclear deployment scenario from Chapter 22, it becomes energy-independent of every global supply chain. The Strait of Hormuz ceases to matter. OPEC ceases to matter. Russian gas ceases to matter. The hemisphere runs on domestic oil and gas in the transition period and nuclear plus renewables in the long term. This is the strongest leg of the Western Hemisphere model, and it's the one that makes everything else possible.

Food. The hemisphere produces more food than it consumes and exports the surplus. The fertilizer supply chain is the vulnerability: natural gas for Haber-Bosch is domestic, but phosphate comes primarily from Morocco and potash partly from Russia and Belarus. The fix is obvious: develop Western Hemisphere phosphate deposits, which exist in smaller but usable quantities, and rely on Canadian potash, which is abundant. The genetic engineering path from Chapter 25, nitrogen-fixing cereals, would eliminate the nitrogen dependency entirely. The food model is resilient, not perfect, but resilient with targeted investment.

Demographics. The hemisphere faces the global fertility decline but with better tools to manage it: immigration, AI-driven productivity enhancement, longevity science extending productive lifespans, and a younger population base in Latin America. The dependency ratio pressure is real but arrives later and less severely than in China or Europe. Manageable with the tools in the toolkit.

Climate. Here the model gets complicated. The western US faces severe drought. The Gulf Coast and eastern seaboard face hurricane intensification and sea-level rise. Central America faces heat and water stress that will drive migration northward. The Amazon is under threat from deforestation and climate change, which could convert it from a carbon sink to a carbon source, with cascading effects on the regional climate. None of these are civilization-ending for the hemisphere. All of them require massive investment in adaptation. The climate introversion problem from Chapter 15 applies: resources spent on adaptation are resources not spent on competition.

Military. The hemisphere is geographically defensible to a degree that no other region can match. Two oceans. Two weak borders. No peer military competitor within 5,000 miles. The Monroe Doctrine, updated for the 21st century, is enforceable without the global force projection that the current strategy requires. A hemispheric Navy doesn't need to patrol the Indian Ocean or the South China Sea. It needs to control the Atlantic and Pacific approaches and the Caribbean. That's a smaller, more sustainable mission than global hegemony.

Innovation. The hemisphere retains the world's strongest innovation engine if the US maintains its research funding, immigration, and university system. The threats described in Chapter 27 are real but self-inflicted, which means they're fixable. A hemispheric model that includes nearshored manufacturing in Mexico, Canadian resource extraction, and American innovation creates a vertically integrated economy that doesn't depend on Chinese supply chains.

• • •

The Eastern Hemisphere bloc. Start with what China has.

The world's largest population, for now, though it's shrinking. The world's largest manufacturing base by output.

The Belt and Road network spanning Central Asia, Southeast Asia, Africa, and increasingly Europe. Energy access through Russian pipelines and the potential grand bargain for the Far East. Rare earth dominance. Battery manufacturing dominance. A military designed specifically to deny the Western Pacific to American power projection.

Russia as a dependent partner, providing energy, resources, and strategic depth. The Central Asian republics, increasingly integrated into the Chinese economic system. Southeast Asian nations, hedging between China and the US but economically gravitating toward Beijing. Pakistan, a nuclear-armed partner. Iran, a resource-rich client state that has just been weakened by a war that served Chinese intelligence-gathering purposes beautifully.

Africa, where China has built the deepest infrastructure and economic ties of any major power. The Indian Ocean rim, where Chinese-built ports from Djibouti to Gwadar to Hambantota provide a string of naval access points. And potentially, the Arctic, where Chinese-Russian cooperation opens the Northern Sea Route as an alternative to the chokepoints that American naval power currently threatens.

• • •

Stress-test the Eastern bloc.

Demographics. This is China's critical weakness, and the grand bargain doesn't solve it. The Far East provides resources, not people. China's working-age population shrinks by over 200 million by 2050 under official projections, and the real numbers may be worse. The Belt and Road partners provide labor in their own countries but not in China. AI and automation partially compensate, as Chapter 23 described, but the demographic headwind is the single biggest threat to the model. If the workforce decline exceeds the productivity gains from technology, the engine stalls.

Energy. With the Russian Far East and existing pipelines, the Eastern bloc achieves energy security through overland supply chains that can't be interdicted by the US Navy. That's the strategic logic of the grand bargain. The Strait of Malacca vulnerability, which is China's version of the Hormuz problem, becomes manageable if enough energy arrives by pipe and rail. Not eliminated, because oil from the Middle East still matters, but reduced to a level that doesn't threaten regime survival.

Food. This is the second critical weakness. China feeds 1.4 billion people on agricultural land that is being degraded by overuse, contaminated by industrial pollution, and threatened by water scarcity in the north. The warming Russian Far East helps, but developing new agricultural land takes decades. The fertilizer supply chain is partially domestic, partially dependent on imports. The food model is less resilient than the Western Hemisphere's, and the margin for error is thinner.

Climate. China faces severe challenges: water scarcity in the north, flooding in the south, coastal cities at sea-level risk. But it also has advantages: a government that can direct adaptation investment without democratic debate, a construction industry that can build sea walls and relocate populations on command, and a population accustomed to state-directed sacrifice. The climate introversion problem applies, but the authoritarian model may be more efficient at managing it. Maybe. The track record of authoritarian governments managing complex, long-term challenges is mixed at best.

Military. The Eastern bloc doesn't need to project power globally. It needs to dominate the Western Pacific, secure the Belt and Road corridors, and defend the overland supply chains from Russia and Central Asia. That's a regional military mission, not a global one. China's military is designed for exactly this. The A2/AD architecture described in Chapter 4 denies the Western Pacific to American carriers. The ground forces secure borders and Belt and Road nodes. The navy controls the South China Sea and the approaches to the strait. It's a defensible model, and it doesn't require the global reach that is currently bankrupting the US.

Innovation. This is the Eastern bloc's biggest structural disadvantage after demographics. As Chapter 27 argued, China excels at scaling existing technology but struggles at the frontier. In a bipolar world where each hemisphere develops independently, the side that produces more breakthroughs gains a compounding advantage. If the US maintains its innovation lead, the Western bloc's technology base pulls ahead every year, and the gap becomes unbridgeable within a generation.

• • •

The model's stability depends on something that neither side fully controls: the countries in between.

Europe is the biggest question mark. The EU has the world's second-largest economy by some measures, a highly educated workforce, sophisticated manufacturing, and cultural ties to the United States. In the bipolar model, Europe naturally aligns with the Western Hemisphere bloc. But Europe is also geographically part of Eurasia, dependent on energy imports that have historically come from Russia and the Middle East, and connected to China through trade volumes that make decoupling economically painful.

A Europe that aligns firmly with the Western Hemisphere strengthens the bloc enormously: adding 450 million educated people, a massive economy, and manufacturing capability that complements American innovation and Latin American resources. A Europe that hedges, trying to maintain economic ties with both blocs, weakens both and creates a buffer zone that is politically unstable and strategically unreliable. A Europe that fragments, with individual nations making separate deals with each bloc, produces chaos that benefits neither.

The early signals are mixed. European defense spending has surged since Russia's invasion of Ukraine, with several NATO members finally hitting the 2 percent of GDP target they had been dodging for decades. The EU passed the AI Act in 2024, asserting regulatory leadership over a domain the US has left ungoverned. Britain's AUKUS partnership with the US and Australia signaled a post-Brexit turn toward the Anglosphere and the Pacific. But Orban's Hungary is functionally a Russian client state inside the EU. The AfD in Germany and Le Pen's movement in France pull toward nationalism and away from Atlantic solidarity. European energy dependence on Russian gas, which the Ukraine war was supposed to end, has been replaced by dependence on American LNG and Middle Eastern supply chains that run through the same chokepoints this book has been describing.

The nuclear dimension adds another layer. American B61 tactical nuclear weapons are stationed in Belgium, Germany, Italy, the Netherlands, and Turkey under NATO's nuclear sharing arrangements. If Europe fragments or hedges, those weapons become a liability rather than a deterrent: nuclear warheads hosted by countries whose commitment to the alliance is wavering, in a political environment where a Le Pen or an AfD chancellor might demand their removal. The nuclear sharing framework was built for a unified NATO facing a single threat. It was not designed for a contested Europe where some members are drifting toward neutrality while hosting American nuclear weapons on their soil. Europe is not automatically in the Western bloc. It is being contested, and the outcome depends on whether Washington offers a partnership worth having or drives Europe toward the hedging strategy that serves nobody.

India is the other great swing state, and it is complex enough to require its own chapter.

The Middle East, Africa, and Southeast Asia are the contested periphery. Each region will be pulled by both blocs, offered investment and security in exchange for alignment. The countries that make the best deals will prosper. The countries that choose poorly, or can't choose at all, will become the spaces between, the ungoverned zones and conflict areas where the competition's friction generates heat.

• • •

The bipolar model is not elegant. It's not utopian. It doesn't solve the problems described in this book so much as it provides a framework for managing them. Two blocs, each internally coherent enough to address their own crises, each strong enough to deter the other from military adventurism, each focused on their own hemisphere rather than trying to dominate the entire planet.

The model breaks in four scenarios, and honesty requires naming them. First, if the economic interdependence between the blocs is too deep to sever. The semiconductor supply chain runs through both hemispheres. Rare earth processing is dominated by China. American companies manufacture in Chinese factories and sell to Chinese consumers. Decoupling sounds clean in a policy paper. In practice it means rewiring supply chains that took thirty years to build, at a cost of trillions and a decade of disruption. If the decoupling fails or stalls, the bipolar model collapses into a messy hybrid where economic integration persists even as strategic competition intensifies, which is roughly the world we have now and the world that produced the current crises.

Second, if the Western Hemisphere can't hold together politically. The model assumes hemispheric unity under American leadership, but Latin American countries have their own interests, their own grievances with Washington, and their own Chinese investment to consider. Brazil, the hemisphere's second-largest economy, has been deepening ties with China, not pulling away. Mexico is a manufacturing platform for both blocs. A Western Hemisphere that fragments along political lines is not a bloc. It's a neighborhood.

Third, if the climate transition accelerates faster than either bloc can manage, producing the fortress-state fragmentation scenario from Chapter 15 where every country turns inward and the concept of hemispheric cooperation becomes meaningless.

Fourth, if the Eastern bloc fractures from within. The book has presented China as a patient strategist executing a multi-decade plan, and in many respects that description is accurate. But patience is easier to sustain when the economy is growing. China's property sector has wiped out trillions in household wealth. Belt and Road lending has collapsed by roughly two-thirds since 2020 as recipient countries default and Beijing pulls back. Youth unemployment is running at levels the government stopped reporting because the numbers were embarrassing. Local government debt is a ticking bomb. And Xi Jinping has purged his own military leadership so aggressively that the PLA's senior officer corps is selected for loyalty rather than competence, with no succession plan for a 72-year-old leader who has abolished every norm that would produce one.

A China that implodes economically, or that enters a chaotic succession crisis, does not neatly fit the bipolar model. It produces a power vacuum in the Eastern Hemisphere that could be more dangerous than Chinese dominance, because a fragmenting great power with nuclear weapons and territorial disputes on every border is not a stable entity. The model assumes two coherent blocs. If China can't hold its bloc together, the model collapses into multipolarity, which is harder to manage than bipolarity and historically more prone to war.

These are real risks. They could happen. The model survives them only if the Western bloc invests in the internal cohesion, the infrastructure, and the institutional capacity described in Part Six. Without those investments, the bipolar model is a plan on paper and the actual outcome is chaos.

The interdependence objection deserves a direct response, because it is the strongest argument against the model. The US and China trade over $700 billion annually. Supply chains in electronics, pharmaceuticals, and rare earths run through both countries. Chinese holdings of US Treasury debt, while declining, remain substantial. Economists who study trade networks argue that this level of integration makes clean decoupling impossible without catastrophic economic disruption on both sides. They are right about the clean part. Clean decoupling is a fantasy. But the bipolar model does not require clean decoupling. It requires strategic decoupling in the sectors that matter for security: semiconductors, rare earths, defense components, pharmaceutical precursors, critical minerals. The CHIPS Act, the export controls on advanced chips to China, and the European "de-risking" strategy are all partial decoupling measures already in motion. The goal is not zero trade between the blocs. The goal is reducing dependency in critical sectors to the point where neither side can weaponize a supply chain during a crisis. That is achievable. It is already happening, slowly and unevenly. The bipolar model requires it to happen faster and more deliberately.

It's the least bad option. And in a world of bad options, the least bad is the best available.

The question that determines whether the least bad option is good enough is the one the next chapter asks: if you had to choose, which hemisphere would you rather live in? Not which one wins. Which one provides a better life for the people who live under its system.

Because in the end, the purpose of power isn't power. It's the kind of life the power makes possible.

Chapter 30: The India Question

India is the other swing state. And India is more complicated than Europe because India is not a finished product. It is a civilization in the middle of the most consequential transformation any country has attempted since China's reform era began in 1978.

Start with the demographics, because they explain everything else. India surpassed China as the world's most populous country in 2023. Its 1.46 billion people have a median age of 29, compared to China's 40. Sixty-eight percent of Indians are of working age. By 2030, one out of every five working-age people on the planet will be Indian. This is the demographic dividend that every economist talks about, the massive young labor force that can power decades of growth if the economy creates enough jobs to absorb it. The window is open now and it closes by the mid-2040s, when India's working-age population peaks and the dependency ratio starts climbing.

But here is the part most Western analysts miss: India's fertility rate has already dropped below replacement. The national total fertility rate hit 2.0 in 2021 and is still falling. Southern Indian states like Kerala, Tamil Nadu, and Andhra Pradesh have fertility rates between 1.4 and 1.6, comparable to Western Europe. The northern states, Bihar, Uttar Pradesh, Jharkhand, still have higher rates but are converging fast. India is not going to be an endlessly growing population. It is going to peak at roughly 1.7 billion around mid-century and then start declining, following the same trajectory as every other country that has industrialized and educated its women. The question is whether India can convert its demographic dividend into economic growth before the window closes, or whether it ages before it gets rich, which is the nightmare scenario that China is already living.

The conversion requires institutional capacity that India is still building. The economy is the fifth largest in the world at roughly $3.5 trillion and projected to pass Germany and Japan by 2030. But the growth is concentrated in services, particularly IT, where five million workers in Bangalore and Hyderabad serve the global technology industry. Manufacturing, the sector that lifted China's hundreds of millions out of poverty, is still only about 15 percent of India's GDP compared to China's 28 percent. Modi's Make in India campaign has been ambitious in rhetoric and modest in delivery.

The infrastructure gap is a generation wide: Indian roads, ports, and railways lag China's by twenty years of investment. The education system produces world-class engineers from the Indian Institutes of Technology but fails hundreds of millions of rural students who attend schools with absent teachers, no books, and collapsing buildings. The caste system, officially abolished but socially entrenched, remains a structural drag on human capital development, locking talent out of opportunities based on birth rather than ability.

India has the raw material to be a great power. Whether it has the institutional capacity to convert raw material into strategic weight within the demographic window is the open question of the century.

The governance problem is distinct from the economic one and in some ways harder. India is the world's largest democracy, which means it has the self-correction mechanism that Chapter 31 argues is the Western model's decisive advantage. But Indian democracy also produces coalition governments that struggle to sustain policy over multiple election cycles, a bureaucracy that the British designed for control rather than development and that independent India never fully reformed, and a federal structure where state governments vary wildly in competence and corruption. Gujarat and Tamil Nadu run like mid-tier developed countries. Bihar and Uttar Pradesh, which together contain more people than Brazil, run like failed states. The national average conceals two Indias that coexist uneasily within the same borders.

Modi's government has pushed centralization and digital infrastructure aggressively. The Aadhaar biometric identity system covers 1.3 billion people and has enabled direct cash transfers that bypass corrupt local officials. The Unified Payments Interface processes over 10 billion transactions per month, more than Visa and Mastercard combined in India. The digital public infrastructure is genuinely world-class and has no equivalent in any other developing country.

But digital infrastructure does not build roads, ports, or factories. It does not train welders or nurses. It does not fix the schools in rural Bihar where a child's odds of learning to read depend more on which state she was born in than on any national policy. India's digital leap has created a visible modernity that coexists with a physical infrastructure that cannot support the weight of the ambition.

Pakistan is the structural drain that no amount of economic reform can eliminate. India spends roughly 2.4 percent of GDP on defense, which sounds modest until you calculate what that buys: the fourth-largest military budget on Earth, consumed disproportionately by the need to garrison a disputed border with a nuclear-armed neighbor that sponsors cross-border terrorism and has fought four wars against India since 1947. Every dollar spent defending against Pakistan is a dollar not spent on infrastructure, education, or the industrial development that would convert demographic potential into economic power. China has no equivalent drain. Japan has no equivalent drain. America has not shared a hostile border with a nuclear-armed adversary since the Cold War ended. India lives with this permanently, and it shapes every strategic calculation the country makes. Operation Sindoor in May 2025 demonstrated both India's military capability and its constraint: the strikes were precise, the S-400 performed, the Pakistani response was contained, but the crisis consumed diplomatic bandwidth and defense resources that India would rather have directed toward the Chinese border or economic development. Pakistan is the weight that slows India's rise, and neither country can remove it.

Some analysts argue that India will not remain a swing state. They argue it will become a third pole: large enough, nuclear-armed enough, and economically significant enough to operate independently of both the American and Chinese systems. The model has historical precedent. During the Cold War, India led the Non-Aligned Movement, which was a third position between the American and Soviet blocs. Modi's multi-alignment is the updated version. What differs is scale. Cold War India was poor and strategically marginal. A 2040 India with a $7 to $10 trillion economy, 200 million more working-age people than China, a blue-water navy, and nuclear weapons capable of reaching Beijing and Washington is not marginal. It is a pole. Whether the bipolar model can accommodate a third pole without collapsing into the messier multipolarity this book has warned about depends on whether India's interests align closely enough with either bloc to function as a de facto partner, even without formal alignment. The bet this book makes is that China's border aggression and India's democratic values push it toward the Western camp in practice, even if Delhi never says so publicly. That bet could be wrong. If it is, the bipolar model weakens considerably.

The military picture is formidable on paper. India has the fourth-largest defense budget in the world, roughly $75 billion in 2026. It has an estimated 160 or more nuclear warheads with a no-first-use policy. Its army is one of the largest on Earth. Its navy operates aircraft carriers, nuclear-powered submarines leased from Russia, and a growing fleet of indigenous warships. The BrahMos supersonic cruise missile, a joint venture with Russia, is one of the few Indian defense exports with genuine international demand. The Philippines bought it. Vietnam and Indonesia are in negotiations.

But India's military has a dependency problem that the 2025 conflict with Pakistan exposed. Roughly 60 percent of India's major weapons platforms are of Russian origin. Its front-line fighters are Russian Su-30MKIs. Its main battle tanks are Russian T-72s and T-90s. Its most capable air defense system is the Russian S-400, five batteries worth $5.4 billion, which proved its value during Operation Sindoor by downing Pakistani aircraft and drones. After the conflict, India immediately moved to buy 288 additional S-400 missiles from Russia for $1.1 billion to replenish what it had used. The world's largest arms importer is still buying from a country under comprehensive Western sanctions. India walks a diplomatic tightrope that would snap under the weight of a less important country.

Washington has granted India a CAATSA waiver for the S-400 purchases, citing the broader strategic interest of keeping India in the American orbit as a counterweight to China. That calculation tells you everything about India's position. It is too big to sanction, too important to ignore, and too independent to control. India buys Russian weapons, buys discounted Russian oil in defiance of Western sanctions, maintains a close diplomatic relationship with Moscow, and at the same time deepens its security cooperation with the United States, Japan, and Australia through the Quad framework. It bought French Rafale fighters and is negotiating for 114 more. It operates American P-8I maritime surveillance aircraft. Its trade with the United States reached $132 billion in fiscal year 2025, making America India's largest trading partner for the fourth consecutive year. Its trade with China, despite the border tensions, was $128 billion in the same period.

The China friction is real and structural, not performative. In June 2020, Chinese and Indian soldiers fought a brutal hand-to-hand battle in the Galwan Valley along the disputed border in Ladakh, the first lethal clash between the two countries in 45 years. Twenty Indian soldiers and an unknown number of Chinese soldiers were killed. Both sides have since deployed tens of thousands of troops and heavy equipment to the Line of Actual Control, the undefined border that stretches across some of the most forbidding terrain on Earth. The two largest countries in the world, sharing a 2,100-mile border, have no agreed boundary. That dispute is not going to be resolved. It is going to be managed, sometimes well, sometimes badly, for the rest of this century.

India's energy vulnerability connects directly to the Hormuz chapter. India imports over 85 percent of its oil, much of it from the Persian Gulf through the Strait of Hormuz. When the strait closed during the Iran war, India was hit hard. Unlike China, which had stockpiled oil in advance, India lacked the strategic reserves to cushion the shock. India is also one of the countries most vulnerable to climate change. Groundwater depletion across the northern breadbasket states threatens agricultural productivity. Heat waves that were once rare are now killing thousands annually. Monsoon patterns are becoming less predictable, which matters enormously for a country where agriculture still employs 46 percent of the workforce. The Himalayan glaciers that feed India's major rivers are retreating, which means the rivers that water northern India's farms will carry less volume in the decades ahead.

So where does India land in the bipolar model? The honest answer is: wherever India decides, and India will decide based on who offers the better deal. India has been non-aligned since independence in 1947, and its current government, under Narendra Modi, has rebranded that tradition as "multi-alignment": taking the best offer from everyone, committing fully to no one. India will buy American drones and Russian air defense systems. It will join American-led naval exercises and buy discounted Russian oil. It will condemn Chinese border aggression and maintain $128 billion in trade with

Beijing. This is not hypocrisy. It is the rational strategy of a country that is powerful enough to have options but not yet powerful enough to dictate terms.

A non-aligned India, playing both sides for maximum advantage, is the most likely near-term outcome. A firmly aligned India, in either direction, changes the model substantially. India in the Western bloc adds 1.4 billion people, a growing economy, a nuclear arsenal, and the world's largest young workforce to the American side of the ledger. India in the Chinese bloc, which is far less likely given the border tensions, gives Beijing demographic ballast it desperately needs and a southern flank that completes the encirclement of the Indian Ocean. The competition for India's alignment will be one of the defining diplomatic contests of the next thirty years, and it will be won not with weapons or ideology but with investment, technology transfer, energy partnerships, and respect for India's insistence on strategic autonomy.

Chapter 31: The Better Overlord

If you had to choose, right now, which hemisphere to live in for the rest of your life and the lives of your children and grandchildren, which one would you pick?

Not which one is stronger. Not which one wins the competition. Which one provides a better life for the people who live under its system. Because that's the question that matters. Power is a means, not an end. The end is the quality of life that power makes possible. And the two hemispheres offer very different answers to that question.

• • •

Start with what the Chinese model provides. Credit where it's due: the Chinese Communist Party presided over the most dramatic reduction in poverty in human history. Eight hundred million people lifted out of extreme poverty in forty years. Cities built from nothing. Infrastructure that makes American infrastructure look like it was built by amateurs, because in many cases it was. High-speed rail that connects the country. A manufacturing base that produces everything from iPhones to aircraft carriers. Rising living standards for hundreds of millions of people who, a generation ago, were subsistence farmers.

That's real. It's impressive. And it came at a price that most Americans wouldn't pay.

Social credit systems that monitor behavior and restrict access to transportation, education, and employment based on compliance scores. A surveillance infrastructure of cameras, facial recognition, and AI monitoring that tracks citizens in every public space. Internet censorship that blocks access to information the government deems threatening, which includes most of what you're reading right now. The imprisonment of over a million Uyghurs in Xinjiang in what the international community has documented as a campaign of cultural genocide. The crushing of democratic institutions in Hong Kong. The suppression of any political organization that challenges Party authority. Disappearances. Forced confessions. Labor camps.

This isn't Western propaganda. It's documented by organizations across the political spectrum, confirmed by satellite imagery, leaked government documents, and the testimony of survivors. The Chinese model produces material prosperity and political repression at the same time, and the Party's position is that the trade-off is worth it. Stability and growth in exchange for freedom and autonomy.

For the billions of people who would live under Chinese influence in the Eastern Hemisphere, the model is clear: your material needs will be met, possibly very well, as long as you don't challenge the people meeting them. The trains will run on time. The buildings will be modern. The internet will be fast, within the boundaries the government sets. Your children will be educated, in the subjects and the worldview the government approves. And if you step out of line, the system will find you, and the consequences will be swift and invisible.

• • •

The American model provides something different. And the honest version of "something different" has to include the failures alongside the strengths, because this book has not been shy about the failures.

The American model provides political freedom that is messy, inefficient, frequently infuriating, and occasionally captures its own institutions for the benefit of the people running them rather than the people they're supposed to serve. Chapter 20's account of private equity extraction is a catalog of American institutional failure. Chapter 21's account of the education system is a catalog of American social failure. The Iran war from Part Two is a catalog of American political failure. This book has not been kind to American performance, because American performance has earned the criticism.

But here's what the American model provides that the Chinese model doesn't: the ability to criticize the system from within and change it.

You can write a book called *The Enshittification of America* and publish it without going to prison. You can call the President a fool on national television and not disappear. You can organize a political movement to change the policies you disagree with, and while the system will fight you with money and inertia and institutional resistance, it won't send the secret police. You can read this book, which criticizes every major American institution from the military to Congress to Wall Street, and the worst consequence is that someone on social media might call you a name.

That's not nothing. That's everything.

The capacity for self-correction is the structural advantage that doesn't show up in GDP figures or military budgets. China's system optimizes brilliantly for the priorities the leadership sets. If the leadership is right, the system outperforms every democracy. If the leadership is wrong, there is no mechanism for correction short of catastrophe. Mao's Great Leap Forward killed 30 to 45 million people because nobody could tell Mao he was wrong. Xi's zero-COVID policy locked a billion people in their homes for three years because nobody could tell Xi it wasn't working. The one-child policy created the demographic crisis that is now China's greatest strategic vulnerability because nobody could challenge the policy until the damage was irreversible.

Democracy is slow. It's wasteful. It makes stupid decisions because voters are sometimes stupid and politicians are often worse. But it has a built-in error-correction mechanism that no authoritarian system possesses: when things go wrong, the people in charge can be removed without a revolution. When policies fail, they can be changed without a palace coup. When institutions are captured by extraction interests, reformers can fight back through legislation, litigation, and public pressure.

The reform may be slow. It may be incomplete. It may come too late in specific cases. But the mechanism exists, and its existence means that the failures described in this book are temporary rather than permanent. Private equity can be regulated. The education system can be reformed. Military procurement can be restructured. Immigration policy can be changed. None of these reforms are easy, but all of them are possible within the system.

Under the Chinese model, the equivalent reforms happen only if the paramount leader decides they should happen. If he doesn't, they don't. And there is no mechanism to replace a paramount leader who is wrong, short of his death or a power struggle within the Party. The system has no bottom-up correction. It has top-down direction, which works when the direction is correct and produces catastrophe when it isn't.

Xi Jinping has made this structural flaw worse, not better. The CCP managed two orderly power transitions after Mao: Jiang Zemin to Hu Jintao, and Hu to Xi. Both followed informal norms of term limits, age-based retirement, and the grooming of successors years in advance. Xi abolished all three. He removed presidential term limits in 2018, packed the Politburo Standing Committee with loyalists at the 2022 Party Congress, and has groomed no successor. He is 72 years old. There is no heir apparent. There is no faction with the institutional base to manage a transition. And Xi has systematically eliminated anyone who might have served in that role.

Since 2023, he has purged his own foreign minister, two consecutive defense ministers, the commander of the Rocket Force and most of its senior staff, both vice chairmen of the Central Military Commission, and dozens of generals across every branch of the PLA. At least 642,000 Party cadres at various levels were punished in the first three quarters of 2024 alone. These were not rivals or opponents. They were Xi's own appointees, people he personally promoted, removed because they built the kind of independent networks that any successor would need to govern.

What results is a system where no one with talent or ambition dares to develop an independent power base, because having one is grounds for investigation. The bench is empty by design. When Xi dies or becomes incapacitated, the world's second-largest economy and third-largest nuclear arsenal will have no prepared leader, no institutional process for selecting one, and a senior military whose officer corps has been gutted and rebuilt for loyalty rather than competence. The CCP has been through this before. It was called the Cultural Revolution. The aftermath took a decade to stabilize.

The honest counterargument is that the self-correction mechanism is under attack and may be degrading. The same openness that allows democratic societies to criticize their governments and change course also makes them vulnerable to information warfare that corrupts the mechanism from the outside. Chapter 11 described the foreign information campaigns that flood democratic information environments with contradictory narratives until the cost of determining what is true exceeds what most citizens will pay. The goal is not to convince people of a lie. The goal is to destroy their confidence that truth is knowable at all. If that confidence breaks, the self-correction mechanism breaks with it, because self-correction requires a population capable of distinguishing good policy from bad, competent leaders from corrupt ones, real threats from manufactured ones. A population that has given up on the possibility of knowing what is real cannot self-correct. It can only react to whatever narrative reaches it loudest and most often. Russia, China, and Iran all understand this, and they are investing heavily in making it happen.

The question for the Western model is whether the self-correction mechanism is robust enough to survive the attack on it, or whether the openness that makes democracy work is the very vulnerability that destroys it. The answer depends on whether democratic societies can defend the information environment the way they defend physical territory: as a domain that requires active protection, investment, and institutional capacity.

So far, the response has been inadequate. Social media platforms built business models that reward outrage and polarization. Governments have been slow to regulate. Citizens have been left to handle an information environment designed to manipulate them, with no institutional support comparable to what authoritarian citizens receive from their filtered, controlled, but at least internally coherent information systems. Fixing this is not optional. It is a precondition for every other reform in this book.

• • •

The innovation dimension connects directly to this. Chapter 27 argued that breakthrough innovation requires intellectual freedom: the ability to challenge orthodoxy, pursue unfashionable ideas, publish uncomfortable results, and fail repeatedly without punishment. Democratic societies provide this, imperfectly but structurally. Authoritarian societies suppress it, because the same intellectual freedom that produces technological breakthroughs also produces political dissent, and the system can't have one without risking the other.

Over the next fifty years, the technologies described in Part Six will determine which hemisphere pulls ahead. AI, nuclear energy, longevity science, genetic engineering, human enhancement, each requires the kind of breakthrough thinking that freedom enables and control suppresses. The Western Hemisphere's innovation advantage is not an accident. It's a structural consequence of political freedom, and it compounds across every technology domain at once.

China will continue to excel at application, scaling, and deployment. It will build more solar panels, more batteries, more high-speed rail, more of everything that can be manufactured by following a plan. But the breakthroughs that change the plan, the ones that create entirely new technological domains and render existing competition obsolete, will come disproportionately from the hemisphere where people are free to think dangerous thoughts.

The economic freedom dimension is less dramatic but equally consequential. In the Western model, you can start a business without government permission. You can own property with legal protections that survive changes in political leadership. You can accumulate wealth without worrying that a policy shift will confiscate it. You can sue the government and sometimes win. These are not abstractions. They are the conditions that make long-term investment rational. A Chinese entrepreneur who builds a successful technology company can have it restructured, nationalized, or its founder disappeared if the Party decides the company has become too powerful or too politically inconvenient.

Jack Ma, once the richest man in China and the founder of Alibaba, criticized financial regulators in 2020 and vanished from public life for months. His Ant Group IPO, which would have been the largest in history at $37 billion, was cancelled by government order. The message was received by every entrepreneur in China: the Party giveth, and the Party taketh away.

Western economies have their own problems with regulatory capture, crony capitalism, and the PE extraction this book has documented at length. But the baseline property rights, the rule of law, the independent judiciary, the constitutional protections against arbitrary seizure, create an environment where building something for the long term is a rational bet. In the Chinese system, the long-term bet always carries the risk that the rules change without warning and without recourse.

• • •

The social dimension matters for the people living inside the system, and it matters for the system's ability to attract talent from outside.

Chapter 12 established immigration as the critical demographic advantage. The United States attracts immigrants because it offers something that no amount of Chinese economic growth can replicate: the possibility of becoming American. The cultural identity of the United States, however imperfect and contested, is at its core open. You can come from anywhere, speak any language natively, practice any religion or none, and become a citizen with the same legal rights as someone whose family has been here for ten generations. No other major country on Earth offers this, and China specifically does not.

China is an ethnostate. Not officially, but functionally. Han Chinese identity is the organizing principle of the society. Non-Han minorities are tolerated in theory and suppressed in practice, as the Uyghurs and Tibetans can attest. Immigration to China is negligible. Naturalization is nearly impossible. A brilliant AI researcher from Nigeria or Brazil or India can move to the United States, get a green card, start a company, and become an American. The same researcher cannot become Chinese in any meaningful sense. China will never be an immigration destination the way America is, and that means China cannot solve its demographic crisis the way America can.

The talent magnet effect compounds everything else. The best researchers, entrepreneurs, and engineers from around the world migrate toward the system that lets them in, lets them build, and lets them become part of the society. The hemisphere that attracts the most talent produces the most innovation, which produces the most prosperity, which attracts more talent. The virtuous cycle runs on openness, and openness is something the Chinese system is structurally incapable of providing.

. . .

None of this is to say that the American-led hemisphere is good. It's to say that it's better. And "better" is a low bar when the comparison is a surveillance state that puts ethnic minorities in camps.

The American model has enormous problems. This book has spent several hundred pages documenting them. The military is overextended and under-resourced. The political system is dysfunctional. The industrial base has been hollowed out by financial extraction. The education system is failing. The infrastructure is crumbling. The healthcare system is a disaster. The wealth inequality is obscene. The political discourse is poisoned. The Iran war is a strategic blunder that is consuming resources needed elsewhere.

All true. All problems that the system has the tools to fix if the political will materializes. And the political will is more likely to materialize in a system where the public can demand it than in one where public demands get you monitored, scored, and silenced.

The optimistic case for the Western Hemisphere isn't that it's currently performing well. It's that it has the structural capacity to perform better. The innovation engine. The immigration magnet. The error-correction mechanism. The resource base. The geographic advantage. The tools described in Part Six. The capacity to learn from failure, which is the one capacity this book is betting the entire argument on.

The pessimistic case for the Eastern Hemisphere isn't that it's currently weak. It's that its strengths are brittle. The demographic decline that no amount of state direction can reverse. The innovation ceiling that no amount of investment can break without freedom. The dependency relationships that work only as long as Beijing can afford to maintain them. The absence of error-correction that means a single bad decision at the top can cascade through the entire system without a brake.

● ● ●

The people in the middle, the billions in Africa, South Asia, Southeast Asia, the Middle East, and parts of Europe who will live in the contested space between the two hemispheres, face a choice that is not really a choice. They will align with whichever power offers the better deal, and "better deal" will mean different things in different places. For some, it will mean infrastructure investment. For others, security guarantees. For others, market access. For others, simply being left alone.

The Western Hemisphere's pitch to these countries has historically been democracy and human rights, which sounds good in a speech and carries very little weight in a conversation about port construction or debt relief. The Eastern Hemisphere's pitch is roads, railroads, and no lectures about your internal affairs. The second pitch is winning in most of the contested world, and it will continue to win unless the Western bloc matches it with a competing economic offer rather than moral exhortation.

The better overlord isn't the one with better values. It's the one that delivers better outcomes while maintaining enough freedom for the people under its umbrella to fix the things that don't work. That's the Western bet. It hasn't been paying off lately. But the capacity is there. And the alternative is a system where the capacity for improvement depends on the judgment of one man and the obedience of a billion people who have no mechanism to tell him he's wrong.

If you're picking a hemisphere for your grandchildren, pick the one where your grandchildren can complain about it. That's the one that gets better over time.

Chapter 32: The Migration Century

In 2015, roughly a million refugees crossed the Mediterranean into Europe. Most were fleeing the Syrian civil war, which was itself partly triggered by the worst drought in the region's recorded history. The political impact of that million people on European politics was seismic. Brexit. The rise of far-right parties across the continent. The near-collapse of the European Union's migration policy. Viktor Orbán in Hungary. Marine Le Pen in France. The AfD in Germany. A million people reshaped the politics of a continent of 450 million.

Now multiply that by a hundred. Because that's what's coming.

. . .

The forces described in this book don't just create crises. They create movement. When land stops growing food, people move. When water runs out, people move. When coasts flood, people move. When heat becomes lethal, people move. When wars destroy cities, people move. When economies collapse, people move. Every force in every chapter of this book produces, as one of its downstream effects, a human being picking up whatever they can carry and walking toward somewhere better.

The World Bank estimates that by 2050, climate change alone could displace 216 million people within their own countries. The actual number of people who cross international borders will be a fraction of that, but a fraction of 216 million is still tens of millions. Add the displacement from conflicts, food crises, water scarcity, and economic collapse, and the total number of people in motion during the next three decades is measured in the hundreds of millions.

This is not a refugee crisis. A crisis implies something temporary, an emergency that is dealt with and then subsides. This is a structural condition. The 21st century will be defined by the permanent, accelerating movement of people from places that are becoming uninhabitable to places that aren't. The migration will not stop because the forces driving it will not stop. The climate will keep warming. The water will keep declining. The food will keep getting more expensive. The populations in the most vulnerable regions will keep growing, at least for another generation, before the global fertility decline reaches them.

The countries that figure out how to manage this movement will thrive. The countries that try to stop it will fail. And the countries that pretend it isn't happening will be overwhelmed.

• • •

The directions of flow are predictable because geography and economics dictate them.

Sub-Saharan Africa to North Africa and Europe. The Sahel is expanding. Lake Chad is disappearing. Agricultural land is degrading. The population is the youngest and fastest-growing on the planet. The infrastructure described in Chapter 13 doesn't connect African economies to each other, so the path of opportunity leads north and then across the Mediterranean. Europe is aging, shrinking, and in desperate need of working-age people. The supply of migrants and the demand for labor should be a natural match. Instead, European politics treats every arriving boat as an invasion.

Central America and the Caribbean to North America. The Northern Triangle countries, Guatemala, Honduras, El Salvador, face a combination of violence, corruption, climate stress, and economic stagnation that has been driving migration northward for decades. Climate change intensifies every one of those factors. The Mexican border will be the pressure point for as long as the United States remains the most attractive destination in the hemisphere, which is to say forever.

South and Southeast Asia to wherever will take them. Bangladesh facing inundation. Vietnam's Mekong Delta flooding. Myanmar's displaced Rohingya. Climate refugees from the Pacific islands. India's internal displacement from drought and flooding. Numbers are enormous, the destinations are uncertain, and the political capacity of the receiving countries is inadequate.

Middle East and Central Asia outward. The water crises described in Chapter 15. Iran's collapsing agricultural system, now compounded by war damage. Iraq's declining rivers. Afghanistan's perpetual dysfunction. The flows go toward Europe, toward the Gulf states, and toward anywhere the labor market can absorb them.

• • •

Here is where this chapter connects to the rest of the book's argument, and where the politics get difficult.

Chapter 12 established that the United States needs immigration to solve its demographic problem. Not wants. Needs. Without immigration, the US workforce shrinks, the tax base erodes, the innovation pipeline thins, the military can't recruit, and the structural advantages described in every subsequent chapter weaken. Immigration is not a policy option. It's a survival strategy.

Chapter 31 established that the ability to absorb immigrants is one of the Western Hemisphere's decisive advantages over the Eastern one. China can't do it. Japan can't do it. South Korea can't do it. The United States can, and has been doing it for 250 years. The talent magnet effect is the virtuous cycle that feeds innovation, which feeds prosperity, which feeds the talent magnet. Break the cycle and the hemisphere's structural advantage disappears.

And yet the dominant political discourse in the United States treats immigration as a threat rather than a lifeline. The border wall. The family separations. The deportation campaigns. The rhetoric about invasion and replacement and cultural dilution. Every election cycle produces a bidding war over who can sound tougher on immigration, and the policy moves steadily toward restriction regardless of which party holds power.

The people driving this discourse are, with very few exceptions, not thinking about demographics. They're not thinking about innovation pipelines. They're not thinking about workforce ratios or tax bases or military recruitment. They're thinking about the cultural anxiety that a million people crossing the Mediterranean produced in Europe, and they're projecting it onto the American context. The fear is real. The policy response to the fear is suicidal.

• • •

The smart version of immigration policy, the one that serves the strategic interests this book has described, looks nothing like what either party is currently proposing.

First, separate the flows. Economic migration and refugee flows are different phenomena with different causes and different solutions, but the current system treats them identically: everyone is either legal or illegal, and the political debate is about the wall between the two categories. A functional system would create multiple pathways: a fast-track for high-skill workers, a managed program for seasonal agricultural labor, a refugee processing system that operates offshore rather than at the border, and a regularization process for the millions of undocumented workers who are already embedded in the economy.

Second, invest in the source countries. The cheapest way to reduce migration pressure is to make the source countries livable. If Guatemalan farmers can grow food, they don't walk to Texas. If Senegalese graduates can find jobs, they don't cross the Sahara. The billions spent on border enforcement and detention would produce more results if spent on agricultural development, water infrastructure, and economic opportunity in the countries people are leaving. This is not charity. It's strategic investment in reducing a problem at its source rather than trying to contain it at the destination.

Third, make immigration part of the hemispheric strategy. The Western Hemisphere model described in Chapter 29 requires labor mobility within the bloc. Mexican workers in American factories. Colombian engineers in Canadian mines. Brazilian agricultural workers on Argentine farms. The hemisphere's labor force is its labor force, regardless of which side of which border individual workers were born on. A hemispheric labor market, with managed flows and mutual recognition of credentials, would be the economic equivalent of the military alliance: a shared resource that makes the whole stronger than the sum of its parts.

Fourth, and this is the one that requires the most political courage, tell the public the truth. The United States needs immigrants the way it needs water. Not because immigrants are charity cases who deserve sympathy. Because they are the workforce, the taxpayers, the innovators, and the military recruits that the country's survival depends on. Every economic model, every demographic projection, every military recruitment analysis says the same thing: without immigration, the United States declines. With it, the United States remains the most competitive nation on the planet for at least another century.

The politician who says this clearly, with the data and the conviction to make it stick, will be doing the most strategically important thing any American leader has done since Truman decided to contain the Soviet Union. The politician who builds a wall instead is condemning the country to the same demographic fate as Japan, except without Japan's cultural cohesion or technological discipline.

• • •

The migration century will be disruptive regardless of how well it's managed. Hundreds of millions of people in motion produce friction. Cultural friction. Economic friction. Political friction. Communities that have been stable for generations find themselves changed by arrivals who speak different languages, practice different customs, and compete for housing and jobs. The friction is real. Pretending it doesn't exist is as foolish as pretending migration can be stopped.

The question is whether the friction is managed or whether it's exploited. Politicians who exploit migration anxiety for votes are making the problem worse while claiming to solve it. A wall doesn't stop a force that is being driven by the four horsemen described in Part Four. It just channels the force into more dangerous pathways: desert crossings, smuggling networks, undocumented populations that live in the shadows and can't be integrated because the system refuses to acknowledge they exist.

The countries that manage migration successfully will be the ones that treat it as what it is: a structural force that can't be stopped but can be directed. Water doesn't stop flowing because you build a dam. It flows around the dam, or over it, or through it. The smart response is to build channels that direct the water where you want it to go. The migration century requires channels, not dams.

The brain drain argument is the one that makes people in developing countries angry, and they have a point. When the best-educated doctor in a Kenyan hospital emigrates to London, Kenya loses a decade of training investment and a community loses its healthcare. When an Indian engineer graduates from IIT and moves to Silicon Valley, India subsidized the education and America harvests the innovation. The sending countries bear the cost. The receiving countries capture the value. This is extraction, and the parallels to the PE model are not accidental: the system is structured to move value from places that can least afford to lose it to places that benefit most from gaining it.

But the brain drain argument has a flip side that the data increasingly supports. Remittances, the money immigrants send home, totaled over $650 billion in 2023, more than three times global foreign aid. That money goes directly to families, bypassing the corruption and overhead that swallow most government-to-government aid. Diaspora networks connect sending countries to global markets, capital, and knowledge in ways that no aid program replicates. Indian engineers in Silicon Valley didn't just leave India. They created the networks that produced India's $250 billion IT industry. Chinese researchers who trained in American labs brought expertise home that accelerated China's technology sector by decades. The brain drain is real. The brain circulation that follows it is also real, and over time the circulation can return more value to the sending country than the original departure cost. The question is not whether talented people leave. They will, because opportunity is unevenly distributed and always will be. The question is whether the systems that receive them and the systems that lose them are designed to convert the movement into mutual benefit rather than one-way extraction.

History supports this. The last great age of migration, from roughly 1880 to 1920, moved tens of millions of people from Southern and Eastern Europe to the Americas. The United

States absorbed over 20 million immigrants in those four decades. The receiving societies were transformed, sometimes painfully, always permanently. The political backlash produced the Immigration Act of 1924, which virtually shut the door for forty years. The economic and strategic consequences of that closure are still debated, but the demographic contribution of the earlier wave is not. The industrial workforce that built the Arsenal of Democracy in World War II was overwhelmingly the children and grandchildren of those immigrants. The scientists who built the atomic bomb included refugees from fascism who came because America would take them when nobody else would. The engineers who built the space program, the entrepreneurs who built Silicon Valley, the doctors and nurses who staff American hospitals: the lineage traces back, again and again, to someone who showed up with nothing and built something.

The migration century ahead will be larger in absolute numbers, more diverse in origin, and more complicated in its political dynamics. But the fundamental economic logic is the same. People move from places with fewer opportunities to places with more. The places that receive them get workers, taxpayers, consumers, and entrepreneurs. The places that reject them get older, smaller, and weaker. The demographic math from Chapter 12 is not optional. Countries that close their borders to migration while their fertility rates sit below replacement are choosing to decline. They may decline with cultural purity intact, if that matters to them. But they will decline.

• • •

This chapter sits near the end of the book because migration is where every other force converges. Demographic decline creates the demand for migrants. Climate change creates the supply. Food and water crises accelerate the supply. Conflicts multiply the supply. Economic extraction, the Enshittification pattern, creates the housing costs and wage pressures that make migration politically toxic in the destination countries. The education crisis means the native workforce can't fill the roles that the economy needs, increasing the demand for immigrant labor. The innovation pipeline depends on immigrant talent. The military depends on immigrant recruits.

Pull on any thread in this book and it leads to a person walking toward a border.

The last chapter pulls all the threads together. It's the one this book is named for. Because when you've mapped the forces, identified the tools, stress-tested the model, and confronted the migration reality, the question that remains is the simplest and the hardest: what do you do about it?

You're stuck in the middle. Clowns to the left of you. Jokers to the right. And the only way forward is through.

Chapter 33: Clowns to the Left, Jokers to the Right

You've made it through 32 chapters. You now know more about the structural forces shaping the next fifty years than most people in Congress, most pundits on television, and most of the talking heads who get paid to have opinions about the world. You know about the missile math, the shrinking fleet, the wars without reasons. You know about the four horsemen. You know about the quiet empire and the grand bargain. You know about the tools that could fix it and the extraction that's preventing the fix. You know about the two hemispheres and the migration century and the better overlord.

Here is what I actually think happens. Geopolitics tells us America goes through cycles of engagement and withdrawal. We are in a withdrawal phase now. We continue pulling inward for another decade or so, fighting lots of small wars but avoiding the big one, letting allies figure out more of their own problems, paying the price in influence and credibility. Then something shifts, the way it always shifts, and we come back with a roar. We have done this before. The difference this time is the internal damage to our institutions, our industrial base, and our capacity to think clearly about what we face. That is what this book is really about. Not whether America survives. It will. But whether it comes back capable of the role it has to play.

And if you're like most readers at this point, you're feeling two things at the same time. First, a certain grim clarity about how bad the situation is. Second, a certain frustrated anger that the people in charge don't seem to know any of this or care.

That's the stuck-in-the-middle feeling. That's the title of this book. You see the problems. You see the solutions. And you're stuck between a political left that fights cultural battles while the material foundation erodes and a political right that sells nationalist grievance while handing structural advantages to competitors. Clowns to the left of you. Jokers to the right.

This chapter is about what you do with that feeling. Because the feeling, by itself, is useless. It's worse than useless. It's paralyzing. And paralysis is the one thing this moment doesn't allow.

. . .

Let me be blunt about the political situation, because the book has earned the right to be blunt.

The American left, broadly defined, has spent the last two decades focused on cultural issues that are real but not existential. Representation matters. Inclusion matters. The dignity of marginalized communities matters. None of those things build a nuclear reactor. None of them weld a submarine hull. None of them produce a bushel of wheat or a THAAD interceptor or a longevity drug. The left has captured the language of progress while abandoning the material conditions that make progress possible. You cannot run a civilization on good intentions. You run it on energy, food, manufacturing, and the educated workforce that produces all three.

The left's failure is a failure of priorities. The issues it focuses on are not wrong. They are small relative to the forces bearing down on the country. Debating the language of inclusion while China builds railroads across three continents is a failure of scale. The left has made itself the party of symbolic progress while abandoning the material conditions that make progress possible.

The American right, broadly defined, has spent the same two decades building a rage machine that produces votes and nothing else. The grievance politics, the culture war, the performative cruelty toward immigrants, the war on expertise, the reflexive opposition to any government action that isn't military spending or tax cuts. The right has captured the language of strength while pursuing policies that make the country weaker. Cutting research funding weakens innovation. Restricting immigration weakens demographics. Gutting environmental regulation doesn't strengthen the economy. It degrades the resource base the economy depends on. Defunding education doesn't save money. It produces the deskilled nation described in Chapter 21.

The right's failure is a failure of seriousness. The anger it channels is real. The solutions it offers are fake. Restricting immigration doesn't solve the demographic crisis. Cutting taxes doesn't rebuild the industrial base. Pulling out of international agreements doesn't make the country stronger. Worshiping the military while hollowing out the defense industrial base is a form of hypocrisy that every chapter of Part One documented.

Both sides are failing because both sides are fighting the last war. The left is fighting the cultural battles of the 2010s. The right is fighting the economic battles of the 1980s. Neither is fighting the structural battles of the 2030s, because the structural battles require an understanding of the forces described in this book, and those forces don't fit on a bumper sticker or in a campaign ad. The failures are not equivalent in kind. The left's failure is neglect: ignoring the material foundations while arguing about representation. Neglect can be corrected by shifting priorities. The right's failure is active destruction: defunding research agencies, attacking universities, dismantling the regulatory frameworks that the solutions in Part Six require. Destruction has to be rebuilt from rubble. The distinction matters, even though both outcomes leave the country unprepared for what is coming.

• • •

The optimistic thesis of this book, which I have maintained through several hundred pages of documentation about how badly things are going, is simple: every problem described here is solvable. Not easily. Not quickly. Not without cost. But solvable.

The missile production bottleneck is solvable with sustained industrial investment and workforce development. The shipyard crisis is solvable with a national vocational training strategy and a twenty-year shipbuilding commitment. The Iran war's strategic drain is solvable by ending the war and redeploying the assets to the Pacific. The demographic cliff is solvable with immigration and longevity science. The food crisis is solvable with nuclear hydrogen and precision agriculture and eventually nitrogen-fixing cereals. The climate crisis is solvable with nuclear energy and adaptation investment. The pandemic vulnerability is solvable with mRNA platforms and AI surveillance and genetic resistance. The PE extraction is solvable with tax reform and regulatory oversight. The education crisis is solvable by paying teachers, rebuilding vocational training, and deploying AI tutoring. The innovation pipeline is solvable by funding it.

Every single one of these solutions exists. They are not theoretical. They are not waiting for a breakthrough. They are sitting on the shelf, developed and demonstrated, waiting for someone to decide to deploy them. The obstacle is not technology. It is not money, though money is required. It is not knowledge, though knowledge helps. The obstacle is political will, which is another way of saying it's us.

To be fair, some tools are being picked up. The CHIPS and Science Act committed $52 billion to domestic semiconductor manufacturing. TSMC is building a fab in Arizona. Samsung is building one in Texas. Intel broke ground in Ohio. The

Inflation Reduction Act directed hundreds of billions toward clean energy manufacturing. These are real investments. They are also a fraction of what China has deployed over the past decade, they depend on sustained political commitment that could evaporate with the next administration, and the timelines for producing actual chips from the Arizona fab stretch into the late 2020s. The reshoring movement is real. It is not yet at the scale or speed that the competition demands. Acknowledging progress is not the same as declaring victory. The gap between what is being done and what needs to be done remains enormous.

. . .

The constraint is always political will. That's the sentence that echoes through every chapter, and it's the sentence that this closing chapter has to confront honestly.

Political will, in a democracy, comes from the public. Politicians do what the public demands, or at least what the public will tolerate. If the public demands action on the defense industrial base, action happens. If the public demands immigration reform, reform happens. If the public demands nuclear energy deployment, reactors get built. The machinery of democracy is clunky and slow and corrupted by money, but it responds to sustained public pressure because politicians want to keep their jobs.

The problem is that sustained public pressure requires an informed public. And an informed public requires the very institutions this book has documented being destroyed: a functioning education system that produces citizens capable of understanding complex issues, and a functioning media that explains those issues in terms that citizens can act on. When the education system produces people who can't read a newspaper article and the media that would have written the article has been gutted by private equity, the feedback loop that democracy depends on breaks.

This is where *The Death of Thinking*, the first book in this series, connects to everything else. The cognitive decline is not a separate problem. It's the meta-problem. It's the reason the other problems aren't being solved. A population that can't think critically can't evaluate politicians, can't understand policy trade-offs, can't distinguish between solutions and scams, and can't sustain the attention required to hold leaders accountable over the multi-year timelines that serious problems require. The death of thinking is the death of self-governance, and the death of self-governance is the death of everything.

And *Turn Off the TV, Get Off Your Ass, and Do Something* is the action plan. It's the book that says: the institutions are broken, but you're not, and the act of engaging, demanding, organizing, and refusing to accept the unacceptable is the mechanism that makes democracy work even when the institutions are failing. One person doesn't change the system. A million people, informed and angry and organized, change it overnight. That's what the civil rights movement demonstrated. That's what the labor movement demonstrated. That's what every successful reform movement in American history demonstrated. It starts with people who refuse to accept that the way things are is the way things have to be.

. . .

Here is what I'm asking you to take away from this book.

The forces are real. Demographic decline, resource depletion, climate change, pandemic risk, great power competition, and the wars that flow from all of them are not hypotheticals. They are happening now. They will intensify. Pretending they don't exist or that the market will sort them out or that God will provide or that some politician will fix it is not a strategy. It's surrender.

The tools are real. Nuclear energy solves energy, food, and water at the same time. AI multiplies the effectiveness of everything else. Longevity science rewrites the demographic model. Genetic engineering breaks the food chain dependency and creates pandemic resilience. Human enhancement creates a qualitatively more capable civilization. The innovation engine that produces these tools is the single most important strategic asset the country possesses. Every one of these tools exists and works. None of them are deployed at the scale the moment demands.

The obstacle is us. Not technology. Not money. Not knowledge. Us. Our failure to demand that our political system treat these issues with the seriousness they deserve. Our tolerance for politicians who offer cultural warfare instead of infrastructure. Our willingness to be distracted by the latest outrage cycle instead of focusing on the structural forces that will determine whether our grandchildren live in the most powerful civilization on Earth or in a declining ex-superpower that traded its future for tax cuts and Twitter fights.

The window is finite. China's demographic clock gives us maybe 15 to 20 years before their window for expansion closes. The climate transition is already underway and accelerating. The nuclear technology is available now but takes a decade to deploy at scale. The AI competition is happening in real time. The Iran war is consuming the military capacity we need for everything else. Every year of delay is a year of advantage handed to competitors who are not delaying.

So here is what Monday morning looks like. Not for the president or Congress or the Pentagon. For you. The person holding this book.

Know what your representatives are doing. Not what they say on cable news. What they vote for. What they fund. What they cut. Every member of Congress has a voting record. Every committee hearing is public. The organizations that track this are not hard to find. The Federation of American Scientists tracks defense and nuclear policy. The Center for Strategic and International Studies publishes accessible analysis of every strategic issue in this book. The Congressional Budget Office publishes cost estimates for every major piece of legislation. The Government Accountability Office audits federal programs and publishes the results. These are not partisan organizations. They are the institutional infrastructure of democratic accountability, and they work only if citizens use them.

Demand specific things. Not "fix the economy" or "be tough on China." Specific things. Fund NRC reform so nuclear licensing takes three years, not fifteen. Fully fund the TAME trial so the first anti-aging drug reaches the public instead of languishing in a funding gap. Double the DARPA budget, which costs less than a single aircraft carrier. Create a fast-track immigration pathway for STEM PhDs who want to stay after graduating from American universities. Mandate that the Pentagon audit itself and pass, something it has failed to do eight consecutive times. Require PE firms to disclose what happens to defense contractors after acquisition. Fund vocational training at the scale the shipyard crisis demands. These are not left-wing or right-wing positions. They are structural investments that benefit every American regardless of which cable news channel they watch.

Run for something. School board. City council. County commission. State legislature. The people making the decisions that shape your community are not smarter than you. Many of them are less informed than you are now, having finished this book. The structural problems described in these pages play out locally: the school that doesn't teach kids to read, the zoning board that blocks the housing development, the county commission that rejects the solar farm or the nuclear plant. National policy matters, but local policy is where most Americans encounter the consequences of institutional failure, and it is where a single informed person can make the most difference.

Talk to people who disagree with you. Not to argue. To listen. The polarization described in these pages is real, but it is sustained by isolation, not by irreconcilable values. Most Americans, when you get them away from the tribal signaling of social media, agree on more than they disagree on. They want their kids to get a good education. They want to be able to afford a house. They want the roads fixed and the water clean and the country safe. The disagreement is about means, not ends, and the means cannot be debated productively if the two sides never talk to each other. Every conversation with someone from the other tribe that doesn't end in a screaming match is a small structural repair to the social fabric.

Support the institutions that produce real information. Subscribe to a newspaper. Fund investigative journalism. The Alden Global Capital story from Chapter 20 is a story about what happens when the institutions that hold power accountable are destroyed. Democracy requires informed citizens, and informed citizens require reporters who do the work of gathering facts and holding officials to account. The local paper that covers your school board, your city council, your county government, is the last line of defense between you and officials who would rather operate in the dark. If that paper dies, the dark wins. This is not an abstraction. It is a measurable, documented phenomenon: corruption increases in communities that lose local news coverage. Paying for journalism is a civic act.

$$\cdot \ \ \cdot \ \ \cdot$$

I started this book with a photograph of the USS Missouri. Nine sixteen-inch guns. A floating fortress that could kill anything in line of sight. And then I showed you the USS Zumwalt: smooth, sleek, and functionally unarmed because the Navy traded mass for precision and ran out of ammunition when the math turned against it.

That image is the metaphor for the country. America traded its manufacturing base for financial engineering. Traded its education system for diploma mills. Traded its infrastructure for tax cuts. Traded its shipyards for stock buybacks. Traded its strategic reserves for quarterly earnings reports. Traded mass for precision everywhere, in every domain, and is now discovering what happens when a country optimized for efficiency encounters a world that demands resilience.

But the Missouri had something the Zumwalt didn't, and it wasn't just the guns. It had 2,700 people. Americans who showed up, did the work, maintained the ship, fired the weapons, and won the war. The hardware mattered, but the people mattered more. The hardware could be replaced. The people were the irreplaceable part.

They still are.

The structural forces in this book are impersonal. Demographics don't care about your politics. Climate doesn't care about your feelings. Resource depletion doesn't care about your ideology. The forces will do what forces do. The question is whether the people who live in the path of those forces will organize, demand, build, and adapt, or whether they'll sit on the couch watching cable news and wonder why nobody is doing anything.

You're stuck in the middle. The left won't help you because it's arguing about the wrong things. The right won't help you because it's selling you fake solutions to real problems. The institutions that are supposed to represent you have been captured by interests that don't care about you. The media that's supposed to inform you has been hollowed out by extraction. The education system that's supposed to prepare you has been degraded into irrelevance.

You have this book. You have the information. You know more about the actual shape of the world than the vast majority of your fellow citizens. What you do with that knowledge is the only variable in this story that I can't predict, because it depends on you.

The tools exist. The problems are solvable. The window is open. The only thing missing is the will to act.

So. What are you going to do about it?

Conclusion: The View from Here

I started writing this book in the fourth week of the Iran war, with oil at $107 and climbing, the USS Gerald R. Ford in Crete getting its laundry room fixed, China sending another intelligence vessel to watch the show, and the President on Truth Social claiming victory while 53 percent of Americans disagreed. I finished it after the April 7 ceasefire, with Khamenei dead, his son Mojtaba installed as Supreme Leader by the IRGC, and the Gulf states calling Beijing instead of Washington. The war I opened the book worrying about ended before I closed it, and it ended worse than the worst case I had sketched.

By the time you read this, more of those facts will have changed. The ceasefire may hold or it may have collapsed. Oil may have stabilized or it may have spiked again. Mojtaba may be consolidating power or fighting to keep it. New crises will have arrived that I can't predict. That's the nature of writing about a moving target.

But the structural arguments in this book won't have changed, because structural forces move slowly enough to be predictable and fast enough to be consequential. The Navy will still be too small. The industrial base will still be hollowed out. China will still be building railroads. The demographics will still be declining. The climate will still be warming. The fertilizer will still depend on natural gas. The next pandemic will still be incubating somewhere.

And the tools will still be sitting on the shelf.

• • •

When I look at the world described in this book, I see two futures, and the distance between them is the width of a decision.

In one future, the United States continues on its current trajectory. Fighting wars it doesn't need to fight. Ignoring the industrial base. Restricting immigration. Defunding research. Letting private equity strip the institutions the country depends on. Staring at Taiwan while China builds an empire. Arguing about culture while the material foundations of civilization erode. In this future, the US declines the way every overextended empire declines: not in a dramatic collapse but in a slow, grinding loss of capacity that the people living through it barely notice until it's too late. The frog in the pot.

In the other future, someone decides to act. Not a president, though a president would help. Someone. A movement. A generation. A critical mass of people who understand what's at stake and refuse to accept the current trajectory. In this future, the nuclear reactors get built. The immigration system gets reformed. The education system gets rebuilt. The PE extraction gets regulated. The research gets funded. The AI gets deployed. The longevity science extends productive lifespans. The genetic engineering breaks the food dependency. The enhancement technology creates a more capable civilization. And the hemisphere that was built for this moment, with the geography, the resources, the people, and the innovation engine, rises to the occasion the way it did in 1941 and 1961 and every other moment when the country decided that the future was worth building.

Picture 2076 in the decline future. An aging population maintained by a healthcare system that can't recruit nurses. Rolling blackouts because the grid was never rebuilt. Chinese-manufactured goods in every store because the domestic factories closed decades ago. A military that deters nothing because the production base behind it was sold for parts. Grandchildren who never learned that the country used to build things, because the schools stopped teaching it and the shipyards stopped doing it. A former superpower that looks a lot like Britain in the 1970s, except without the cultural confidence or the Commonwealth.

Picture 2076 in the other future. A continent powered by a thousand nuclear reactors, energy-independent and food-secure. An AI-augmented workforce where a 75-year-old engineer on longevity treatments has the stamina of a 50-year-old and the experience of a lifetime. Genetic diseases eliminated for every child born in the hemisphere. A space economy mining asteroids for the minerals that used to start wars. A hemisphere that absorbed 200 million immigrants over fifty years and turned them into the workforce, the taxpayers, and the innovators that kept the whole system running. Not utopia. A civilization with problems. But a civilization with the capacity to solve them, because it invested in the tools and the people and the institutions that self-correction requires.

The first future is the default. It's what happens if nothing changes. The second future requires effort, sacrifice, political courage, and sustained attention from a public that has been trained to have the attention span of a goldfish.

I'm betting on the second one. Not because I'm naive. Because the alternative is despair, and I don't have time for despair. Neither do you.

Here is what this book showed that no single chapter could show on its own. The tools don't just coexist. They compound. Nuclear energy powers the data centers that run the AI. The AI accelerates the drug discovery that produces longevity treatments. The longevity treatments keep experienced workers productive for an extra decade, which partially solves the demographic crisis. The demographic breathing room buys time for the genetic engineering to create disease-resistant populations that don't overwhelm the healthcare system during the next pandemic. The reduced pandemic vulnerability frees resources for the infrastructure investment that counters China's Belt and Road. The innovation engine that produces all of it is fed by the immigrants who are attracted by the freedom that the democratic system guarantees. Each tool makes the others more powerful. Each failure to deploy one tool weakens all the others.

I want to be honest about the limits of that claim, because it is the kind of clean machine that ought to make a careful reader suspicious. Real systems do not compound this obediently. The tools also collide. The same nuclear buildout that powers the data centers competes with them for the welders and the electricians and the grid connections, and there are not enough of any of those to go around. The AI that accelerates drug discovery is the same AI that throws people out of the jobs the longevity treatments were supposed to keep them working in. Genetic engineering that breaks the food chain arrives wrapped in a politics ugly enough to stall it for a decade regardless of whether the science works. The immigration that feeds the innovation engine is the single most politically combustible item on the entire list.

Deploy all of this at once and you do not get a frictionless flywheel; you get five enormous national fights happening simultaneously, each one capable of derailing the others. The compounding is real, but it is a best case, not a default. It describes what becomes possible if the country can hold several hard things in its hands at the same time without dropping them, which is precisely the thing it has been worst at. That is the honest version. It is still worth doing, because the alternative is to deploy nothing and lose for certain.

That is why partial solutions are insufficient and piecemeal reform is inadequate. The compound effect is the argument. Deploy the full stack and the problems become manageable. Deploy half of it and the half you skipped drags down the half you built. The adversary understands this. China is deploying nuclear, AI, genetics, space, and infrastructure investment as a coordinated national strategy. The United States is deploying them as disconnected programs competing for budget scraps from a Congress that can't agree on what century it is.

• • •

This series began with the observation that Americans have stopped thinking. It continued with the argument that they need to start doing. It explored the technologies that could augment what they're capable of. It documented the financial extraction that is destroying the institutions they depend on. And it ends here, with the geopolitical context that makes all of it urgent.

The context is this: you are alive at a hinge point in human history. The decisions made in the next fifteen to twenty years will determine whether the 21st century belongs to a civilization that values freedom, creativity, and the dignity of the individual, or to one that values order, obedience, and the power of the state. Both civilizations have strengths. Both have flaws. But only one of them gives you the right to argue about the flaws and the power to fix them.

You live in that one. For now.

The forces described in this book are impersonal. They don't care about your politics, your feelings, your hopes, or your fears. Demographic decline will happen whether you vote Republican or Democrat. Climate change will continue whether you recycle or don't. Resource depletion will accelerate whether you drive a Tesla or a pickup truck. The quiet empire will keep building whether you watch the news or turn it off. These forces are not opinions. They are not predictions. They are the physical, biological, and economic realities of the world you live in.

What is not predetermined is how you respond. That's the variable. That's the only variable that matters. And that's why this book, despite everything it describes, is optimistic.

Because the tools exist. Every one of them. Nuclear energy that solves three crises at once. AI that multiplies everything else. Longevity science that rewrites the demographic model. Genetic engineering that breaks the food chain. Enhancement technology that creates a more capable species. An innovation engine that produces all of it, if it's funded and maintained and pointed at the right problems. A geographic position that is the envy of every nation on Earth. A cultural capacity for self-correction that no authoritarian system can match. And an immigration tradition that provides the one resource no amount of money can buy: people who believe the future is worth the journey.

All of that is yours. All of it is real. All of it works.

The only thing it needs is you.

. . .

I'm going to end this book the way I end every book in this series, because it's the only ending that matters.

You know more now than you did when you started reading. You understand the forces. You see the tools. You recognize the trap and you can see the shape of the exit.

So close this book. Put it down. Stand up.

And do something.

Recommended Reading

The following books, reports, and resources informed the research behind this book and are worth reading for anyone who wants to go deeper on the topics it covers.

Military and Strategy

CSIS, "The First Battle of the Next War," 2023 (Taiwan war game). Congressional Budget Office, annual Navy shipbuilding reports. Bryan Clark, Hudson Institute publications on naval strategy. Congressional Research Service reports on Navy force structure.

Demographics and Resources

Darrell Bricker and John Ibbitson, Empty Planet: The Shock of Global Population Decline. Vaclav Smil, How the World Really Works. Peter Zeihan, The End of the World Is Just the Beginning. Dean Spears and Michael Geruso, After the Spike: Population, Progress, and the Case for People.

China and Geopolitics

Rush Doshi, The Long Game: China's Grand Strategy to Displace American Order. Chris Miller, Chip War. AidData, "Banking on the Belt and Road," research reports. Yi Fuxian, Big Country with an Empty Nest. George Friedman, The Next 100 Years and The Storm Before the Calm. Geopolitical Futures (geopoliticalfutures.com).

Technology

David Sinclair, Lifespan: Why We Age and Why We Don't Have To. Jennifer Doudna and Samuel Sternberg, A Crack in Creation. Kai-Fu Lee, AI Superpowers: China, Silicon Valley, and the New World Order. McKinsey Global Institute, "The Economic Potential of Generative AI," 2023. Daniel Yergin, The New Map: Energy, Climate, and the Clash of Nations.

Private Equity and Institutional Decay

Brendan Ballou, Plunder: Private Equity's Plan to Pillage America. Gretchen Morgenson and Joshua Rosner, These Are the Plunderers. NBER working papers on PE-owned nursing home outcomes. SIPRI reports on private equity in defense supply chains.

Water and Climate

Peter Gleick, The Three Ages of Water. Sandra Postel, Replenish: The Virtuous Cycle of Water and Prosperity. John Fleck, Water Is for Fighting Over: And Other Myths About Water in the West. Bureau of Reclamation, Colorado River Basin water supply and demand studies. IPCC Sixth Assessment Report, Working Group II (Impacts, Adaptation and Vulnerability).

Pandemics and Biosecurity

Michael Osterholm and Mark Olshaker, Deadliest Enemy: Our War Against Killer Germs. David Quammen, Spillover: Animal Infections and the Next Human Pandemic. WHO, Global Antimicrobial Resistance and Use Surveillance System (GLASS) reports. Dan Bebber, "Crop Pests and Pathogens Move Polewards in a Warming World," Nature Climate Change, 2013.

India

Ramachandra Guha, India After Gandhi: The History of the World's Largest Democracy. James Crabtree, The Billionaire Raj: A Journey Through India's New Gilded Age. World Bank, India Development Update reports. Shivshankar Menon, India and Asian Geopolitics: The Past, Present.

Ongoing Resources

Federation of American Scientists (fas.org). Center for Strategic and International Studies (csis.org). Congressional Budget Office (cbo.gov). Government Accountability Office (gao.gov). War on the Rocks (warontherocks.com). USNI News (news.usni.org). World Nuclear Association (world-nuclear.org). World Health Organization (who.int).

ENEMIES OF YOU

About the Author

Richard Lowe has published over 100 books across nonfiction, fiction, and memoir. He is best known for the Enemies of You nonfiction series that traces the structural forces reshaping American life from the individual level to the geopolitical.

The Death of Thinking asked why Americans stopped being able to think critically and what the cognitive decline was doing to self-governance. *Turn Off the TV, Get Off Your Ass, and Do Something* argued that the institutions were broken but the citizens weren't, and laid out the case for direct civic engagement as the mechanism that makes democracy work even when the system is failing. *The Birth of the Augmented Human* explored the technologies that are changing what humans are capable of, from artificial intelligence to genetic engineering to neural interfaces. *The Enshittification of America: How Private Equity Destroyed the Things We Love* documented, industry by industry, how the leveraged buyout model gutted twelve sectors of the American economy. And *Stuck in the Middle* connects all four into a single argument about where the country stands in the global competition that will define the next fifty years.

Each book was written to be read independently. Together, they form an argument that starts with the individual mind, moves through civic participation and technological capability, exposes the financial extraction that is hollowing out the country's institutions, and ends with the geopolitical context that makes all of it urgent.

Lowe's fiction includes *Forty Worlds*, a collection of short fiction that ranges from literary to speculative, and multiple novels. His writing across genres shares a common trait: he refuses to believe that complicated ideas can't be explained to normal people, and he has no patience for books that make simple things sound complicated to impress other writers.

He writes at a ninth-grade reading level on purpose. Not because the ideas are simple. Because the audience deserves clarity.

He lives in the United States.

ENEMIES OF YOU

https://enemiesofyou.com

About This Series

Something is working against you. Not in the abstract. Not against society or the culture or the country in general. Against you, specifically. Your ability to think. Your ability to pay attention. Your ability to understand what's happening in the world and make good decisions about your own life inside it. Your ability to pass something worth having on to the people who come after you.

This series documents what that something is.

Not one thing. Several things, operating at the same time, from different directions, with different tools. Some of them are commercial. Some of them are political. Some of them are foreign. Some of them were designed specifically to do what they're doing and some of them are just the predictable outcome of systems nobody was watching carefully enough. What results is the same regardless of the cause. Something is eating your capacity to think, to participate, to resist, and to build. This series is about what that something is and what you can do about it.

Each book in the series identifies a specific enemy operating against a specific capacity. The Death of Thinking is about what AI dependency does to your mind when you let it think for you. Turn Off the TV is about what passive consumption does to your time and attention when you let platforms have both. The Birth of the Augmented Human is about the path back to your own capability. Stuck in the Middle is about the geopolitical forces reshaping your world without your knowledge or consent. The Enshittification of America is about the financial engineering that stripped the institutions your daily life depended on and left hollow shells in their place. The Emasculation of America is about the deliberate foreign campaign to demoralize and neutralize the men who would otherwise resist. The Villainization of America is about the psychological operation that turned a nation against its own story.

Nineteen books. Nineteen enemies. One argument running through all of them: none of this happened by accident, none of it is inevitable, and all of it can be countered by people who understand what they're dealing with.

You can read them in any order. Each one stands on its own. But if you read them together, something becomes visible that isn't visible in any single book: the pattern. The way cognitive erosion feeds civic collapse. The way civic collapse feeds cultural vulnerability. The way cultural vulnerability feeds foreign exploitation. The way foreign exploitation feeds the economic extraction that makes everything else worse. These aren't separate problems. They're the same problem operating at different scales.

The series is written for normal people living normal lives who suspect that something is wrong but can't quite name what it is. Not for academics. Not for policy people. Not for the already-converted on either side of any political argument. For people who are smart enough to understand the world but haven't been given the information in a form that respects their intelligence without requiring a PhD to decode it.

Every book is written at a ninth-grade reading level. On purpose. Not because the ideas are simple. Because clarity is a form of respect. If you can't explain something clearly, you probably don't understand it yourself.

The series is also optimistic. That will surprise you after a few hundred pages of documented disasters, structural failures, and deliberate attacks. But the optimism is earned, not performed. The tools exist to counter every one of the enemies documented in these books. The examples exist. The knowledge exists. The only thing standing between the current situation and a dramatically better one is the decision to act on what you now understand.

That decision is yours.

Enemies of You Series

The Death of Thinking: The Enslavement of Humanity

A diagnosis of what happens to human cognitive capacity when practitioners consistently outsource the parts of their work that require genuine thinking to AI tools. Not in one session or one project, but across months and years of daily practice that removes the demands that were quietly building something. Following composite characters through the specific moments where the pattern becomes visible, this book traces the mechanisms of cognitive erosion: the convenience trap, the illusion of understanding, the death of the wrong answer, and the transfer of epistemic authority that occurs when humans stop standing outside the AI's framing and examining it.

The Birth of the Augmented Human: The Freeing of Humanity

The companion to The Death of Thinking maps the other path. A notebook before the AI is opened. A paragraph written before the structure is requested. A hypothesis formed before the diagnostic tool is consulted. Small choices in sequence that accumulate, over months and years, into a practitioner who is more capable, more original, and more able to surprise themselves than the practitioner who did not make them. The other path is available. This book is the map.

Turn Off The TV, Get Off Your Ass, and Do Something

Most people complain about not having enough time while spending hours every day staring at screens. This is not an anti-technology book and not a minimalism guide. It is an anti-passivity book built around one specific argument: every platform has a consuming side and a contributing side. The device is identical either way. The relationship to it is not. This book is about crossing that line and what waits on the other side.

Stuck in the Middle: Wars, Weapons, and the Forces That Will Shape the Next Thirty Years

Written against the backdrop of a US-Israel strike on Iran that exposed the hollowness of American military industrial capacity, this book connects cognitive decline, civic collapse, private equity extraction, and great power competition into one argument about where the world is heading. Covering missile math, carrier vulnerability, demographic collapse, the Belt and Road as strategic colonization, and the technologies that could solve every crisis on the horizon, this is the book that ties everything else into one coherent warning. And one earned, hard-won optimism.

The Enshittification of America: How Private Equity Destroyed the Things We Love

A documented investigation into how private equity firms systematically acquired beloved American institutions, loaded them with debt, stripped out everything that made them worth visiting, and walked away wealthy while leaving communities with hollow shells of what they once had.

Airlines. Restaurants. Department stores. Newspapers. Hospitals. Pharmacies. This book names the firms, documents the playbook, and makes the case that the degradation of American commerce was not inevitable. It was deliberate.

The Emasculation of America: How Russia's Long War Against the American Male Is Destroying the Nation From Within

Beginning with a KGB defector's 1984 warning that nobody heeded, this book traces the deliberate Soviet and Russian strategy to defeat America not through military force but through cultural subversion. Seeding an ideology through universities, amplifying it through social media, delivering it through institutions that now enforce it as policy. Applying academic cult identification criteria to gender ideology, documenting the biological attack through endocrine disruption, and tracing China's acceleration of the same strategy through TikTok, this is not a culture war book. It is a national security argument.

The Villainization of America

America ended slavery, defeated fascism twice, rebuilt its enemies after defeating them, created the largest middle class in human history, and produced more medical and technological breakthroughs than any nation that ever existed. Somehow a significant portion of its own citizens have been convinced it is the primary source of evil in the world. This book documents how that happened, who executed it, and why the psychological campaign to make Americans ashamed of their own country is inseparable from the economic and cultural attacks documented in the two preceding volumes.

Watch the Other Hand: Politics as Cover for the Kleptocracy

While Americans argue about culture war flashpoints and election outcomes, a quieter operation has been moving wealth and power from public hands into private ones at a scale most citizens never see. The political theater is real and exhausting and often deeply felt. It is also doing work for the people whose interests would not survive a population paying attention to what was actually happening. This book documents the kleptocratic capture happening behind the visible politics, names the mechanisms, and traces how the visible politics functions to keep attention pointed elsewhere.

Manufactured Fear: How Crisis Becomes Profit

Every era has its emergencies. The current era has manufactured ones, engineered to maintain a state of generalized anxiety that benefits specific industries and political coalitions. The fear is not invented. The proportions are. This book traces how a healthy capacity for legitimate concern was converted into a permanent state of alarm, names the actors who profit from it, and documents what happens to a population that lives at sustained emergency pitch for years on end.

The Death of Privacy: They Know Everything, You Know Nothing

The surveillance system that the citizens of free societies were promised would never be built has been built. Not by a single state with a single agenda but by a coalition of corporate platforms, advertising infrastructure, data brokers, and government agencies that share the substrate even when they do not coordinate the use. This book documents what is actually known about each individual user, who knows it, what they do with it, and what the absence of meaningful privacy means for political freedom in a society that depends on individuals being able to think and act without continuous monitoring.

The Wrong Fight: How the Climate Response Became the Climate Problem

The climate is changing, the consequences are real, and the response that was supposed to address them has been captured by interests that are using the response as a vehicle for their own purposes. The result is a policy regime that produces consequences which would be unacceptable on their own terms but become acceptable because the alternative is framed as denial. This book separates the science from the policy capture, names the specific failures of the current response, and argues for what an honest climate strategy would look like.

The Quiet War: How America's Adversaries Attack Without Firing a Shot

The hot wars of the twentieth century have been substantially replaced, against the United States in particular, by sustained operations that operate below the threshold of military response. Information operations. Cultural subversion. Economic coercion. Cyber penetration of critical infrastructure. Strategic drug supply campaigns. These are the instruments of the quiet war, and they have been working. This book documents the campaigns currently underway against the United States, names the state actors directing them, and explains why the inability to recognize them as warfare is itself one of the campaigns' objectives.

The Dumbing Down: How American Schools Stopped Teaching Children to Think

American schools have been progressively converted from places where children were taught to think into places where children are processed for credentials. The conversion was not an accident or a failure of execution. It was the predictable outcome of policy choices that prioritized measurable outputs over the difficult work of cognitive development, and that defined educational success in ways that did not require it. This book documents what was lost in the conversion, when the choices were made, and what would have to change to teach thinking again.

The Pattern: How the Enemies of You Work Together

The enemies named across this series are not parallel items on a list. They are a system. Cognitive erosion makes civic collapse possible. Civic collapse creates the conditions for kleptocratic capture. Kleptocratic capture funds the manufactured fear that legitimizes the surveillance state. The surveillance state runs on the educational system that produced citizens who cannot evaluate what is being done to them. Each enemy reinforces the others. None of them can be addressed in isolation. This book is the synthesis: how the system operates as a system, why the standard frame of fix-this-one-problem is itself part of the problem, and what counter-strategy looks like for someone who can finally see the whole shape of the attack.

The Debt Trap: How the Financial System Was Designed to Extract From You

Student loans that cannot be discharged in bankruptcy. Credit cards engineered to keep balances revolving. Mortgages structured so the first ten years of payments are mostly interest. Buy-now-pay-later services that have re-engineered impulse purchasing to operate on an installment basis. Auto loans that now run seven years and underwater within twelve months. Each financial product looks like a service. Each one is a specific design choice about who pays whom over time, and the design has consistently moved in the same direction. This book traces the architecture of consumer debt as a wealth extraction system, names the policies and corporate decisions that built it, and explains why the standard personal-responsibility framing is the cover story that lets the system continue.

The Sick Industry: How American Medicine Profits From Keeping You Sick

The American healthcare system spends more per capita than any other developed nation and produces worse outcomes on most measures that matter. The reason is structural. Chronic illness is more profitable than cure. Symptom management is more profitable than prevention. The food industry produces the conditions that the pharmaceutical industry then medicates. The hospital system bills by procedure, not by health. The medical research apparatus is funded primarily by entities with financial interests in particular conclusions. This book documents the architecture of medical extraction, names the specific incentive structures that produce it, and explains why the conversation about fixing healthcare has been confined to the question of who pays rather than what is being paid for.

The Gambling Machine: How America Made Predatory Gambling the Default

In 2018, sports betting was illegal in nearly every U.S. state. By 2024, it was legal and aggressively advertised in most of them. The expansion was not driven by public demand. It was driven by industry lobbying that succeeded because the public attention was on other issues. The new gambling environment is engineered with the full machinery of behavioral psychology: variable rewards, push notifications, free credits that require deposits, in-game betting that runs faster than judgment can keep up with. The financial outcomes are predictable and documented. The social outcomes are accumulating. This book traces how the legalization happened, who profited, and what is now being done to the people the new system has captured.

The Loneliness Engine: How American Life Was Structured to Isolate You

The third places where Americans used to encounter each other are gone. Bowling leagues, fraternal organizations, churches, neighborhood bars, civic clubs, parent-teacher associations: all measurably smaller, in many cases by orders of magnitude, than they were thirty years ago. The replacements are commercial products that provide the appearance of connection while delivering its opposite. This book documents the destruction of the institutions that made American social life functional, names the economic and policy forces that did the destroying, and traces the consequences for mental health, civic participation, and the basic human capacity to be known by other people.

The Theft of Childhood: How American Kids Stopped Becoming Adults

Children spend more time on screens than in any previous generation, less time outdoors than any previous generation, and reach standard milestones of independence later than any previous generation. The teen mental health collapse that accelerated after 2012 is not mysterious. The mechanism is documented. Phone-based childhood, helicopter parenting, the elimination of unsupervised play, the medicalization of normal developmental difficulty, and the school system's drift toward credentials over capacity have produced a generation that is anxious, fragile, and structurally unprepared for adulthood. This book names what was taken, who took it, and what would have to change for the next generation to get a different result.

Books by Richard Lowe

See books by Richard Lowe at

https://masterofworlds.com

Get free publishing insights and industry updates at

https://thewritingking.substack.com

For ghostwriting and book coaching services see

https://thewritingking.com

Index